W9-CPO-510

GOD *According to* GOD

A Physicist Proves We've Been Wrong About God All Along

Gerald L. Schroeder

HarperOne
An Imprint of HarperCollins*Publishers*

 For information address HarperCollins Publishers, 10 East 53rd Street, New York, NY 10022.

HarperCollins books may be purchased for educational, business, or sales promotional use. For information please write: Special Markets Department, HarperCollins Publishers, 10 East 53rd Street, New York, NY 10022.

HarperCollins Web site: http://www.harpercollins.com

FIRST EDITION
Designed by Level C

Library of Congress Cataloging-in-Publication Data are available upon request.

ISBN 978–0–06–171015–5

09 10 11 12 13 RRD(H) 10 9 8 7 6 5 4 3 2

Contents

Introduction

Shortly before sunrise, I stand outside my home in Jerusalem and watch the last moments of night give way to the coming of day. In those early morning hours, with the air spiced with scents from the eucalyptus trees and bushes of thyme that border our courtyard, the sky embraces the fleeting black of night. Pins of starlight mark the grandeur of space. The sun rises and begins to paint the sky blue as the shortest visible wavelengths of the incoming sunlight are scattered across the heavens. The glow of the sky signals the call for prayer.

Jerusalem rests on several hills, and each hillside acts as a reflector, echoing the diverse calls for prayer out again over the city. Today marks the Hebrew month of Elul, the biblical month that precedes the biblical New Year, the holiday of Rosh HaShanah (literally, "the head of the year"). By pleasant coincidence, this year the Muslim month of Ramadan coincides with Elul. Both Elul and Ramadan have special prayers, and that makes this morning's music especially pleasant. Hebrew from a town crier and the blowing of a ram's horn, the shofar, call for Jews to rise and thank God for the magnificent munificence of the day. This mixes with the Arabic from the muezzin asking Muslims to do the same. And then not to be left out of this Divine melody, the bells of the many Jerusalem churches literally chime in, blending perfectly with the voices in Hebrew and Arabic.

Each of our three local cultures yearns to address the one God, Creator of the universe. We may use different languages, but the sense of an underlying Unity remains. This spiritual Oneness, though expressed differently in the three religions, mirrors, as a near replica in the metaphysical realm, the physical unity upon which rest all aspects of the material world.

Much of the four decades of my career as an M.I.T.-trained scientist and, in parallel, the three decades of my study of the Bible has been devoted to probing this physical and spiritual unity. At times the two realms blend, and yet at times they seemed totally and hopelessly at odds. The deeper truth I discovered is that, when we get beyond a superficial understanding of the tangible, material world, we find that the physical and the metaphysical make up a single reality, one world viewed from two vastly different perspectives. It is this that I teach in my classes on science and the Bible.

Albert Einstein discovered that matter is actually pure congealed or condensed energy, energy in the form of solid matter. Everything from our bodies to boulders on a mountain is made of the energy of the big-bang creation. The scientific discoveries of the twentieth and twenty-first centuries have gone a step farther in closing ranks with the creation, finding that matter and the energy from which matter formed are made of something totally ethereal. In physics we call it information or, more extreme, mind. In the words of the knighted mathematician James

Jeans, the world looks more like a great thought than a great machine. Biblical theology agrees totally, telling us, as we will learn, that God used a substrate of wisdom with which to build the world. This Divine wisdom or mind is present in every iota of the world's being. It explains how the energy of the creation, essentially superpowerful light beams, could become alive and sentient, able to feel love and joy and wonder. Divine wisdom was and is present, guiding and forming the way.

The secular world of course takes a different stance. If we can get past the question of what created the universe from nothing (was it God?), we then let the laws of nature take the credit for producing, in some as yet unknown way, the magnificence of life from the big-bang burst of pure, exquisitely hot energy. All this by random chance. It takes a stretch of the imagination, but that is all that is available to a secular explanation of our cosmic genesis.

Though in my books and with my students I present our genesis from a very different view, that of a creating God that is present and active, I too face a dilemma, and my questioning students do not let me ignore the problem. There is something very basic missing in the simplistic view of the God of the Bible operating and controlling the workings of the world. Most obviously, if God is in control, why isn't the world perfect? Not just from our humanly limited view of perfection, but even in a biblical accounting there are multiple examples by which we learn that the world has its faults. Most blatantly, God brought the biblical Flood at the time of Noah to revamp a misdirected world. Couldn't God have foreseen this potential for disaster and nipped it in the bud before it blossomed into a worldwide debacle?

Are we dealing with an absentee God, a God that only once in a while pays attention to the world It created to see if things are going according to some Divine schedule? A superficial reading of the Bible might give that impression. A detailed study of

God as described in the Bible, however, presents a very different picture. For example, as the Israelites are about to enter Canaan, God promises to fight for their victory, but then tells any individuals who have a new home or are recently engaged to marry to return home, lest they die in battle. God promises to fight alongside the Israelites to help gain victory for the army, but there is no guarantee of survival given to any particular individual. In another incident, God promises to send hornets ahead of the Israelite army to drive out the enemy snipers, but not to drive the enemy out too quickly lest the beasts of the field multiply. God could also have controlled the beasts just as God controlled the hornets, but refused to do so. The biblical message is that God is there to help, but steps back, in biblical language hides His face, and insists that we do our part in the job. God has chosen us to be partners.

With the Divine hiding of face, God's presence becomes masked, at times even unpredictable and certainly not always controlling events. This is a dynamic Force, not some static entity able to be pigeonholed into how we think a God should act within Its creation. The overwhelming goodness of the world is so extreme that every sorrow stands out as an unnecessary tragedy. In simplistic terms, God could and should stop every form of undeserved trouble. But that is not the God of the Bible, as the book of Job so blatantly reveals.

The God of the Bible, by the very act of creating the universe, has relinquished a portion of control. With this act, God imbued and empowered humankind with the task of getting a partly perfect world to become fully perfect. This is a tremendous vote of confidence by God in our ability, notwithstanding the fact that God has let us know that we are a stiff-necked and rebellious people. It is as if God has said, "This is what I have to work with, so let's make do with what we've got."

The problem so many people, believers as well as skeptics, have with God really isn't with God. It's with the stunted perception of the biblical God that we imbibe in our youthful years. As children we yearn for a larger-than-life figure who can guide and protect us. Our parents fulfill part of that mission. But the parentlike image of an infinite, error-free God is even more assuring to our young minds. So we grow up retaining this childhood notion of an all-powerful, ever present, ever involved, never erring Creator. Unfortunately, that image fails when as adults we discover that the facts of life are often brutally at odds with this popular, though misguided, piece of wisdom. It's no wonder that atheists chortle at the naiveté of the idea of such a God. We are about to correct that misperception, and in doing so we'll develop an understanding of the Divine as made manifest in our world.

What is the God of the Bible? What can I expect from Him—or Her—or It? What can I demand? Does God want me to make demands? Why did the God of the Bible tell Abraham to sacrifice Isaac, his and Sarah's only child? Does God want us to argue when we confront what appears to be Divine injustice, or are we merely to accept the slap and turn the other cheek? When I feel the surge of emotion at the beauty of a star-studded sky or the joy of a baby's smile, is that a part of the same transcendent God that created a less than perfect world? And if there really is a God, why so often is God's presence so fully hidden that even in the Bible people wonder, "Is there a God among us?" An obvious and predictable God would be so much easier to understand.

By abandoning preconceived notions of the Author of creation and replacing them with the Bible's description and nature's display of God—we will learn about God according to God. The surprise is that the many episodes brought in the Bible mirror with alarming fidelity life as we experience it.

I'm a scientist, and also a student of the Hebrew Bible. The scientific method looks for relationships among seemingly diverse pieces of information, be they held in nature or written in a book. Finding the common ground that binds these sources of knowledge often reveals facts not immediately obvious when considered separately. By combining the information the Bible brings about the nature of God with the discoveries of modern science, I am determined to make sense of why the world runs the way it does, spiritually as well as physically. In this sense I move beyond the scientific interplay between the *Torah* (the Hebrew term for the Five Books of Moses) and *teva* (the Hebrew word for nature) described in my first three books.

This is a search that became for me both academically rational as a scientist, and emotionally spiritual, also as a scientist. The claim in Psalms that "the heavens declare the glory of God and the firmament proclaims His works" (Psalms 19:2) is not a mere metaphor. The study of nature, even with all its intellectual rigor, is filled with spiritual wonder.

ONE

A Few Words About What God Is Not

Before we discuss what God is, it is highly instructive to know what God is not. There is so much misinformation streaming from communities of skeptics and believers alike that separating the Divine wheat from the imagined chaff can be a confusing task.

I am not certain who said that if you can only afford one newspaper, read the opposition. Whoever urged this, it is superb advice. Let's look at what the arguments of bona fide skeptics, purebred materialists, have to teach about our cosmic genesis. As certain as I am that there is a metaphysical dimension active in our world, so they are convinced of the exact opposite, that the world can be described, including its creation, in totally secular terms. They claim that the idea of a Creator, or God, is a human construction, a man-made apparition, arising solely to satisfy an imagined need for an interested cause or force that brought the universe into existence.

The argument against the biblical description of our cosmic genesis is quite basic. If this supposed Creator is actively interested in Its creation, then that Creator has a very perverse sense of compassion and perhaps of humor—more like that of a monster: earthquakes, tsunamis, cyclones have swept hundreds of thousands to their horrendous deaths; approximately eighty million humans have been murdered by fellow humans in the past century. Logic, so their argument goes, dictates that a Creator God, if It existed, would have more empathy in Its guidance of the world It produced.

So powerful is this divergence from the often preconceived notion of how a concerned God "should" behave that Bart Ehrman, chair of the Department of Religion at the University of North Carolina, Chapel Hill, and former pastor of Princeton Baptist Church, rejected his belief in Christianity. The title of his 2008 book provides his reasoning: *God's Problem: How the Bible Fails to Answer Our Most Important Question—Why We Suffer.* "The problem of pain," he writes, "ruined my faith."[1]

Atheists also argue that God is irrelevant, an unnecessary component to human society. Even without a God forcing the rules of the Bible down our throats, humanity would have discovered that communities following the logical laws of society are likely to have a greater chance of survival than loners in the wilderness. This urban style of morality, they reason, merely evolved from—is a more sophisticated version of—our animal ancestors' survival instinct to herd or flock together, as birds of a feather do so well. We don't need a God to tell us that. "Morality," Richard Dawkins, Oxford University professor of the Public Understanding of Science, tells us in his 2007 book *The God Delusion* and again in his BBC documentary *Religion: The Root of All Evil,* "stems from altruistic genes naturally selected in our evolutionary past."[2] For morality we need neither a God nor a Bible.

E. O. Wilson, in his acclaimed book *Consilience,* agrees wholeheartedly. Wilson tells us that "the Enlightenment thinkers . . . got it mostly right the first time. The assumptions they made of a lawful material world, the intrinsic unity of knowledge, and the potential of indefinite human progress are the ones we still take most readily into our hearts." Unfortunately, his "dream of a world made orderly and fulfilling by free intellect" is a dream based on gossamer.[3] It has nothing to do with reality. As Wilson describes in great detail, the intellectual freedom of the Enlightenment itself sowed the seeds for the French revolution's Reign of Terror, in which the leading intellectuals of the day were slaughtered.

In a more recent attempt at achieving the Enlightenment's goal of humankind's free intellect finding the way to peace and fulfillment, we need only turn to the ultimate enactment of the philosophy of Karl Marx, his famous claim that religion is "the opium of the people." And when religion was finally abandoned, we achieved what Marx might have envisioned, had he paid better attention to the lessons of the past. Within a century, Communist Russia (a perversion of the concept of the commune) produced the most uncommunal of all societies, brutal and totally repressive of any form of intellectual freedom.

History repeatedly brings an unwelcome message that we often strive to ignore: the unfettered use of human logic does not lead to a just and moral society, the claims of philosophers Baruch Spinoza and E. O. Wilson notwithstanding. The biological basis of our moral judgments teaches us that the human genome is programmed for pleasure and survival, not for morality.

Of course Dawkins is correct on one point here. Religion is the root of all evil, though not exactly as that statement implies.

The very concept of a definitive evil requires that there be a clear distinction between good and evil. And that distinction is totally of biblical origin. A society based on moral relativity has no fixed bounds. Individuals and groups can decide what is good for them, which in another society or situation might in fact be deemed evil.

When atheists describe God as a sinister monster, a superficial reading of the Bible seems to confirm their view. For starters, we've got biblically condoned slavery (Lev. 25:35–36; Exod. 21:26), genocide in the wars by which the invading tribes of Israel, following the Exodus from Egypt, displace the local tribes of Canaan (Deut. 20:12), and God refusing Moses entry into the Promised Land merely because he made a single mistake (Num. 20:6–12).

Yet the same Book also teaches love of neighbor and the alien (Lev. 19:18, 34). There is one law for both native and foreigner (Lev. 24:22). That demand for equality is extraordinary, especially considering that, amid the grandeur that was Greece and the glory that was Rome, a thousand years after the revelation at Sinai, foreigners were still considered barbarians and were treated likewise. The Torah, the Five Books of Moses (Genesis–Deuteronomy), demands one set of measures and weights for all customers, whether friend or foe (Deut. 25:13), and forbids murder (Exod. 20:13), robbery (Exod. 20:13; Lev. 5:21), oppressing the stranger (an admonition repeated thirty-six times in the Torah), and cruelty to animals (Deut. 25:4). Even the wanton destruction of trees is forbidden (Deut. 20:19).[4] The biblical balance sheet is not as damning as some would have it. But then perhaps the biblical God is schizophrenic—sometimes vicious, sometimes compassionate.

God as schizophrenic? No. But God is also not as simplistic as we often paint God to be. If we take a second look at the Bible, we discover the biblical God reveals a far more complex character than the simplistic version of an always-in-charge, predictable Ruler of the heavens and the earth. When passages of the Bible are quoted out of context, or read in translation, whether that translation is the twenty-two-hundred-year-old Greek Septuagint or a modern English version of the original Hebrew, nuances are often lost. Meanings of words are actually changed to fit within the grammar of the "newer" language.

Certainly if those persons involved in the search for extraterrestrial life received a message from outer space, that message would be studied and analyzed for every nuance. The Bible, if Divine in origin, is a message from totally "outer space." It requires careful study. In this book, we're going to do exactly that.

To take a quick look at some of those passages that atheists and skeptics often bring up, let's consider the nature of slavery as described in the Bible. The Hebrew word for "slave" is "worker," with all the connotations that differentiate the modern concept of slave from that of a worker. In Rome and Greece, the slave was no more than an animated tool. The biblical rules for the humane treatment of slaves are so strict that if a master broke the tooth or any other bodily part of a slave, the slave was immediately freed as a full citizen and with compensation (Exod. 21:26–27). The person about to become a slave must willingly sell him- or herself into slavery (Lev. 25:39, 44), which would only happen due to abject poverty. Kidnapping and selling a person into slavery was a capital offense (Exod. 21:16; Deut. 24:7). Returning an escaped slave to the master was absolutely forbidden (Deut. 23:15–16). A quick reading of *Uncle Tom's Cabin* will put this humane biblical law into juxtaposition with more modern practices of slavery.

As for the genocide found in the Bible, which atheists flag as being condoned by God, we get quite a different perspective when viewed from within the text. The Israelites are about to complete their forty-year trek in the desert, a trek that was to impress upon them a twofold lesson: first, their dependence upon a dependable God if they behaved themselves; and, second, that they possessed the strength and ability to carry out God's plans, that they no longer were a weak and fearful, essentially "institutionalized," group of just-freed slaves. They were a formidable nation. At this point God instructed them to destroy the local tribes that inhabited the land of Canaan. Sounds brutal and would be, if that were the end of God's command. But then we are given the key mitigating piece of information. We are told the reason for this conquest: "In order that they do not teach you to emulate their abominations that they have done for their gods . . . for even their sons and daughters they burn in fire to their gods" (Deut. 20:18; 12:31). When judging the "genocide" accusation, internalize this horrible fact: the Canaanites took their children and burned them to death. But what if they were to abandon their abominations, including child sacrifice? Then the Israelites were to make peace. How do we know this?

When the people of Israel entered Canaan, Joshua was in command. God had instructed Moses to appoint Joshua as leader since he, Moses, was to die prior to their crossing the Jordan River and entering Canaan. Needless to say, the rulers of the tribes of Canaan had no desire to have this new people share their land, especially since this invading people had zero tolerance for one of their cherished customs, the murder of their firstborn boys and girls by burning. And so the locals fought. "And Joshua made war a long time with all these kings. There was not a city that made peace with the children of Israel except the Hivri, the persons of Givon" (Josh. 11:18–19). From this almost seemingly afterthought, that one city made peace, we learn that Joshua offered peace to all the cities. That all but one refused was their choice. Perhaps giving up child sacrifice, if you are addicted to it, is not so simple. Having neighbors who practiced these abominations is socially destructive to the entire society. In offering peace, Joshua was not abrogating God's command. He was merely executing its actual intention, to get rid of the abominations, not necessarily the abominators. To put these acts into perspective, consider living next to a home from which screams of horror and anguish regularly emanate. You discover the cause. Dad is busy raping his daughters while mom gets her pleasure by snuffing out her cigarettes on junior. If you don't take action, then you too are a monster. Now consider discovering that your neighboring village is actively conducting these abominations. That's what Joshua discovered upon his entry to Canaan. I imagine even an avowed atheist steeped in relative morality would recoil at such horror. There was no divine command for genocide; the Canaanites had to either live as decent humans or get out. Had the world taken a lesson from these biblical chapters, Hitler, Pol Pot, and Stalin would have been footnotes to, and not chapters in, history. Being overly righteous and forgiving, reasoning, "After all it's their custom, so why should we impose our values on them?" is no virtue. "There is a time to love and a time to hate; a time for war and a time for peace. . . . Don't be overly righteous or too wise; why destroy yourself" (Eccl. 3:8; 7:16).

Not all of God's traits are friendly. We don't need skeptics to tell us that the God of the Bible wants us to adhere to the standards presented in the Bible. The text itself makes very clear that if we err and diverge from those values there will be a Divine price to pay: "Take heed to yourselves, lest your heart be deceived and you turn and serve other gods and worship them. Then the anger of the Eternal God will be kindled against you and He will shut up the heavens, and there will be no rain and the ground will not yield its fruit; and you shall perish off the good land that the Eternal God gives to you" (Deut. 11:16–17). There seems no room for argument here. And yet we will learn later that in fact God wants us to argue even when we are not totally in the right.

Take for example Moses. Moses, the human with the closest relationship to God, made one error, and God in punishment refused to allow Moses to enter the Promised Land. This, notwithstanding the fact that Moses had led the sometimes quite rebellious people of the Exodus during their forty-year trek in the wilderness prior to reaching Canaan and shaped what was initially a ragtag diverse population into a formidable and unified nation. It took forty difficult years in the desert to achieve this goal, and Moses was up to the demanding task of being the leader the entire time.

As any camper knows, food and water are essentials for any desert trek. And the Exodus was no different. God supplied a daily ration of manna every morning, but water was to be found along the way. And in that naturally arid region water was scarce. At times the supply went dry. At one such juncture, as the people cried for water, Moses grew impatient with them, became angry, and for a brief but fateful moment lost control, in both his actions and his words. "The people complained: 'Why have you brought the congregation of God into the wilderness to die? . . . It is not a place of seeds or figs or vines or pomegranates, and there is no water to drink.' . . . And God said to Moses 'Take the rod, assemble the congregation, you and your brother Aaron, and speak to the rock that it give forth water.' "

Instead of speaking to the rock as God commanded, Moses in anger "said to the people, 'Hear now, you rebels, are we to bring water from this rock?' And Moses raised his hand and smote the rock twice, and water came forth in great abundance. . . . And God said to Moses and Aaron, 'Because you did not believe in me to sanctify me in the eyes of the children of Israel, therefore neither of you shall bring this congregation into the land that I have given to them' " (Num. 20:4–12).

In retrospect Moses's frustration is understandable. Yet the pleading for water by the people was also justified. Moses, in that moment of desperation, had evidenced an expression of doubt in front of the entire congregation: "Are we to bring water from this rock?" Why the "we"? It was God, not Moses or Aaron, who was to bring water from the rock. God had never failed till then and was not about to fail now. But for a moment there was a doubt. Moses later argued with God to rescind the decree and allow him to accompany the people into the land of Canaan, even if he would no longer be the leader. But the damage had been done, and the answer remained no. "And God said, 'Speak no more to Me of this matter'" (Deut. 3:26). In a Divine compromise, God showed Moses the land from the peaks of the adjacent mountains. Then Moses was gathered to his people in peace. God as described in the Bible makes demands and expects us to adhere to those demands. Does that make the image of the biblical God one of a control freak? As we learn more about the nature of Divine control, we'll discover that in most cases there is quite a bit of leeway, a flexible margin, in how we are expected to comply with the rules.

But if God demands our conformance with His commands, can't God also demand the conformance of nature? And if so, then why doesn't God demand that nature behave itself? We are back to Bart Ehrman's query. If God is great, why tragedy? From the aspect of human interactions, induced evil is easily understandable. We have the free will to hurt others. God early in Genesis even admits that, though humans are created in the image of God (1:26–27), "the imagination of humankind's heart is evil from its youth" (8:21). Natural disasters, floods, famines, and earthquakes, however, are another story. Why doesn't the God of the Bible, compassionate by self-definition, control nature? We will explore in this book the weave of human and godly acts that allows such events to happen and our role in partnering with God for their alleviation.

Miracles present a quandary of a different sort. They represent an aberration in the regularity of nature. The regularity of nature is for the most part the sine qua non of the scientific method. And the scientific method is the Holy Grail of skeptics. Miracles are not merely an aberration. They are in the materialist view of the world an imagined perversion of reality.

But if there is a God active in our universe and that God cannot perform miracles, cannot give nature a nudge, then this is not much of a God. Miracles upset the regularity of nature upon which the scientific method of analysis relies. And the scientific method has been phenomenally successful. It has found cures for plagues, sent people to the moon, and lightened the drudgery of the masses. Dr. Norman Geisler, a Christian philosopher and president of Southern Evangelical Seminary, in Charlotte, North Carolina, suggests that there is a way to bring the scientific and the miraculous under a single mantle. His logic is as follows.

We derive laws of nature by observing a consistency between cause and effect for certain categories of events. We derive the laws of gravity by studying the motion of falling bodies, and the laws of thermodynamics by measuring energy transfer in varied reactions, both chemical and physical. But miracles are unique, onetime events. There's no way of correlating cause and effect as a general law for an event that has no repetition. Yet, as Dr. Geisler points out, the logic of scientific scrutiny has been applied to the study of nonrepetitive events.[5]

The lack of repetition has not stopped scientists from speculating on the causes of the two most fundamental phenomena crucial to our existence: the big-bang creation of the universe and the origin of life from nonliving matter. Were these miracles? After all, they were both onetime events.

What caused the big-bang creation? Perhaps there was no cause. Our cosmic origins may derive from some unknowable, uncaused, cosmic fluke. Human logic developed in a world where discovering causes for events was crucial for survival. Was the noise in the tree caused by wind moving the branches or was it the result of a leopard preparing to leap upon one of us? Our cultural history has programmed us to seek causes for all experiences. But perhaps there are uncaused events. In the following two chapters, I discuss in greater depth the implications of uniqueness in several aspects of the physical world. Here we deal with just two onetime events: the origin of the universe and the origin of life.

Though the creation of the universe was an admittedly onetime event, the members of the National Aeronautics and Space Administration (NASA) have attributed a cause to it. NASA in 2006 and 2007 released a vivid diagram depicting the history of our expanding universe. The information illustrated thereon is so significant and fundamental to understanding the history of the universe that the initial research that led to this map earned two persons, John Mather of NASA and George Smoot of the University of California, Berkeley, the Nobel Prize in 2006.

The origin, our cosmic origin, is described in the diagram as the effect of quantum fluctuations in virtual (nonexisting) space. Before the big-bang creation of the physical universe, before the existence of space and energy, quantum fluctuations were possible. This means that, according to this august body of scientists, the laws of nature, or some aspects of the laws of nature—at least quantum mechanics—predate the physical world. If they did not predate the physical universe, then the quantum fluctuations that yielded the universe could not have materialized. The implications of this fact are extraordinary. If time as we understand it is part of the creation, as is usually assumed, then the laws of nature predating the creation must be timeless. They predate time. The laws of nature are totally abstract. They are not nature. They are the laws that will eventually create and govern nature in the universe, once they create the universe by the big bang. This leap of scientific faith is identical to the biblical posit of a timeless, nonphysical Creator, God, having created the universe.

To determine whether timeless laws of nature or a timeless God brought our existence into being, we cannot search into the past and observe the actual act of creation even with the most powerful telescopes. The universe was opaque for its initial three to four hundred thousand years. Our only clues are to be found within the universe the way it is now. Call it a reality check. Does the development of the world from the initial burst of energy that marked the big-bang beginning of our magnificent universe through to the origin of sentient life seem random or guided? If random, then there is no need for a God. But if guided, then clearly we need a Guide. We explore that cosmic evolution and its implications in the next two chapters.

The beginning of life, however, does provide some indication. All forms of life at the molecular, genetic level are so similar that it appears that all life stems from a single origin. Even the simplest of microbes are packed with phenomenally complex and extensive libraries of information encoded within their genetic material. Are the events that led to the onetime origin of life the result of a fluke or is life the result of an information-filled miraculous Cause?

The scientific method observes regularities, repetitions, in nature. If there are no significant exceptions, then logic dictates that these repetitive events represent an underlying law, a system, from which we can generalize how our world functions. We can then use this law to predict events that have not been observed, but are similar in quality to the events of the observed pattern.

For example, day by day we observe that the sun rises in the east. With that repeatability, we've discovered a "law" of nature, the sun rises in the east. With that law we can predict that tomorrow the sun will also rise in the east. Let's relate this concept of observed events and derivation of laws of nature to the origin of the complexity and information sequestered within all forms of life.

Every human experience we have, with no exception, reveals that for ordered complexity to arise and to remain stable, some aspect of the environment must cause its retention. For example, complexity can arise de novo via random acts, as in the shaking of a basket containing many small pieces of paper each with a letter of the alphabet on it. As the letters fall to the basket's bottom, they may land in a way that forms a word. That word would be an example of ordered complexity. But that word is always lost in subsequent tossing of the basket unless the letters that formed the word have been glued in place. Complexity is always lost unless nature somehow makes it last, "glues it in place."

Information may expand and build upon itself, but there is always, in all our experience, an informed or structured basis upon which the expanded knowledge rests and that preserves the information. A self-correcting computer program following specific rules by which it analyzes incoming information and gradually modifies those initially instilled rules is a simple example. The beautiful crystals of salt at the Dead Sea (actually, the biblical name in the original Hebrew is the Salt Sea; Gen. 14:3) are models of repetitive complexity. However, the crystals' formation is anything but random. The ionic or chemical bonding between the atoms of sodium, chlorine, bromine, and a few other elements in the water determines the structure of the crystal. Massive Sequoia trees arise from the information held in a tiny seed. And the first microbe on a formerly lifeless earth? From where did it derive, develop, and retain its wisdom? A "fluke" is not a very scientific way of explaining the phenomena, especially since such a rationalization repudiates all scientific experience.

Harold Morowitz, while professor of molecular biophysics at Yale, computed that to create a bacterium would require more time than the universe has existed, if random combinations of molecules were the only driving force.[6] Physicist Paul Davies was equally intrigued. He believes that life is "built into the scheme of things in a very basic way."[7] Simon Conway Morris, professor of evolutionary paleobiology at the University of Cambridge, finds that "the existence of life on earth appears to be surrounded with improbabilities."[8]

Since life did arise on earth, what was the driving force that provided the information-packed complexity held within even the "simplest" of microbes? The wonder of life is not whether life arose in a microsecond, six days, a billion years, or an eternity. The wonder of life is not how much time passed during which the prebiotic nonliving matter metamorphosed into living cells containing "libraries" of complex information sequestered in their genes. The wonder of life arising from nonliving matter, from rocks and water and a few basic molecules, is life itself. Is life miraculous? According to our understanding of the origins of information and how we make scientific decisions, life's emergence fits the description. Science and miracle in a single sentence. There was a time when that would have been seen as an obvious oxymoron.

The basic question of whether science and religion are mutually exclusive realms reduces to whether there is a place for the metaphysical to be brought within the structure of what until recently was a purely materialist science. The discovery of the big-bang creation of time-space and energy, the metamorphosis of that energy of the creation into particles, and the transformation of those particles into sentient beings, alive with feelings of joy, the transcendental ecstasy of love, and self-awareness, all cry out for an explanation that seems to find its root in something other than the material. The physical particles from which living bodies are constructed, the atoms and molecules, show not a hint of sentience. How can we explain that a bundle of "inert" energy—simplistically stated, superpowerful rays of light—became alive, other than to assume some nonphysical, that is, some metaphysical, input was involved?

The Bible drew this conclusion millennia ago: "I am wisdom. . . . God acquired me [pure metaphysical wisdom] as the beginning of His way, the first of His works of old" (Prov. 8:12, 22).[9] Wisdom, the first of all creations, was and is the driving force behind the sentience of life. Knighted mathematician Sir James Jeans paralleled this insight in his book, *The Mysterious Universe*: "There is a wide measure of agreement, which on the physical side of science approaches almost unanimity, that the stream of knowledge is heading toward a non-mechanical reality. The universe begins to look more like a great thought than a great machine."[10]

And that would answer the ultimate question with which both science and religion struggle. Why is there an "is"? Why is there something rather than nothing? For that answer both science and religion must turn to the metaphysical.

Archskeptic Michael Shermer, atheist par excellence and publisher of *Skeptic Magazine,* in an article in *Scientific American* makes his case for a monistic view of reality: "The body and the soul are the same, and the death of the body . . . spells the end of the soul." There is a millennia-old Talmudic tradition that the role of the Creator will be praised by believers joyfully and by skeptics even against their will. Shermer ends his exposition with "the realization that we exist together for a narrow slice of time . . . , a passing moment on the proscenium of the cosmos."[11] A beautiful choice of words, and oh so true. The *Merriam-Webster*

Dictionary tells us that "proscenium" is none other than the part of the stage in front of the curtain, between the curtain and the audience. Shermer hit the proverbial nail right on the head. We live out our lives on the proscenium, the visible part of the stage. But as every theater devotee knows, the show is directed from behind the curtain.

ONE FINAL NOTE that I know many skeptics will not want to accept. There are several technical claims made in the Bible that appear to contradict firm scientific opinion. The age of the universe, for one. Bible data give a number that is less than six thousand years, while science measures the time since creation in billions of years, some fourteen or fifteen. Were there only six days from creation to Adam? And if Adam and Eve were the first humans, what about all those fossils of hominids dating back seventy thousand years?

As described in detail in my previous books, the ancient biblical commentators, those whose writings predate by many centuries the discoveries of modern science (writers of the Talmud, ca. 400; Rashi, ca. 1090; Maimonides, ca. 1190; Nahmanides, ca. 1250), learned from the detailed wording of Genesis that the universe is young and old simultaneously. These ancient commentators actually discuss what science has only recently discovered, that the flow of time is flexible. The rate at which time passes varies depending upon the conditions and the temporal locations from which events are viewed. That is the nature of time in this amazing world of ours. And with that knowledge they describe the old/young age of our universe. They talk about "beings" that we today would refer to as hominids, beings identical to humans in shape and in intelligence, lacking only the soul of humanity, the *neshama*, to make them human. According to these ancient biblical commentators they walked the earth at the time of biblical Adam and before. "Cavemen" were never a theological problem to these ancient commentators.[12]

A superficial reading of the Bible misses all of this. But the Bible is anything but superficial. The Bible is not an easy read, and nature is not a simple study. Yet as sources of wisdom, for millennia both have intrigued and instructed peoples diverse in race, culture, and homeland. That is because their message is relevant to all times. Our task is to extract the eternal truths held within the "subtext." In the Talmud, a sixteen-hundred-year-old commentary on the Bible, we are told that in the revelation at Sinai the words were written as black fire on white fire. The black and the white are two parts of a single Divine message—the black fire being the recorded text of the Bible, the words that we read, and the white fire being that part of the message subtly held behind the text, sequestered in nature. Only when we understand both the black and the white will we know the full meaning of that evocative message of Sinai. And for that reason the Bible opens with the creation chapter, what the twelfth-century philosopher and theologian Moses Maimonides refers to a *madah teva,* the science of nature. To understand God's actions in this world, Maimonides tells us, we must first understand nature. In the following two chapters we'll look at a few aspects of nature and see if we can discern within *madah teva* the metaphysical hand of God active within the physical world God created.

My hope is that we can leave behind preconceived notions and search for what the world and the Bible actually tell us about the God of creation.

TWO

The Origins of Life

One Reason I Know There Is a God

The most powerful challenge to atheists' view of the world lies within the world itself: the simple reality of existence. Why is there existence? Forget things as complex as life. Just consider the being of anything: space, time, matter in any form. Is there some "law," some axiom, that demands there be existence independent of an underlying force that brought it into being? Even if we posit that the universe and all existence are eternal, the question remains: Why is there an "is"? It's a question that calls out for an answer. Of course the facile response is if there were not existence, then we could not ask the question. True, but we do exist, and so it is a puzzle that demands probing. The greatest self-revelation of a Creator is the creation It brought into being.

Our Cosmic Genesis in a NutShell

big-bang creation of energy ⟶ matter ⟶ life ⟶ brain ⟶ mind and sentience

But how could the energy of the big bang become a sentient mind?

There are two aspects in nature's march toward life that call out for commentary: the creation of a universe perfect for life, and the formation of sentient life able to experience the wonders of love, joy, and compassion, but built of combinations of protons, neutrons, and electrons that have not the vaguest hint of sentience within their structures. Life and consciousness emerged from nonliving matter. How?

I'd studied how the laws of nature make the particles of the world behave. How the elements were formed in stars and supernovae all made and makes logical sense. Granted there is a strong hint of a source of fine-tuning having shaped a world amenable to complex life, but still it was physics. Only when writing my third book, *The Hidden Face of God,* did I truly encounter the hidden hand of God. The majestic subtlety by which the laws of physics gracefully guide the workings of nature to meet the demands of life moves beyond the logical. At the interface between the seemingly inert matter and the life that arose from that inert substrate, for me the metaphysical became apparent.

The basic difficulty in replying to "how" life arose lies within the inherent limits of human brain power. No matter how clever we are or how much advanced education we've absorbed, whether theologian, scientist, philosopher, or New Age guru, we all think within the same box, a box delineated by a logic that can only envision qualities and quantities based on time-space-matter. We can say the word "metaphysical," meaning that which exists outside of the physical, but we cannot comprehend the metaphysical. It simply lies beyond the capacity of the human mind. And if the Bible is correct, then what created our universe, God, was and is metaphysical. How that metaphysical Creator interacts with the physical world need not in any manner conform to how we interact with the world. And so, as the development of life followed a path from the energy of the big bang to the initial microbial life and then to the more complex bodies of animals and the mind of humans, that path may have been directed in ways that are humanly incomprehensible.

In our quest for the origins of life, let's start at the beginning of the evolutionary process. We would do well to ponder, especially in our schools, the puzzle of why there is existence. Unfortunately, by the time we are old enough to even contemplate the wonder of existence, we've been around so long that we just take the fact of existence for granted. But think about it. Why is there anything, why is there a universe within which life may or may not have evolved, developed, rather than nothing? It has been said quite accurately that the difference between nothing (as in before the big-bang creation) and something (the existence of our universe) is infinite.

In a refreshing expression of intellectual honesty, Nobel laureate and theoretical physicist Steven Weinberg, an avid atheist who unabashedly states that "the moral influence of religion has been awful" and further that any "signs of a benevolent designer are pretty well hidden," also tells us that even if we scientists eventually attain a "theory of everything," "we will still be left with the question of 'why?'. . . So there seems to be an irreducible mystery that science will not eliminate."[1]

The conundrum of our cosmic beginning remains. The Bible of course gives God the credit for that event. "God created the heavens and the earth." That is in the very first sentence of the Bible, Genesis 1:1. But the Bible, being God-oriented, has a vested interest in listing God as the creator. Secular science, even as it embraces the concept of a creation, does not necessarily turn to God for the beginning. There are aspects of quantum physics (those are the physical laws that guide the minute subatomic world) that allow the creation of something from nothing. Such a concept seems to violate even a rudimentary understanding of how the world works. But it does not. At the subatomic level something can come from nothing. And our universe may be that something.[2] However, a full exposition on how this can happen is beyond the scope of the current discussion.[3]

But the question is still how? Just what did the big bang produce? Science posits that the big bang was the beginning of time and space. But what about matter? That is considerably more enlightening, literally. The big bang did not produce matter as we know it, not any of the ninety-two elements, such as carbon and oxygen, and not the protons, neutrons, or electrons that would eventually combine to make the atoms of those elements.

By a fraction of a microsecond following the creation, the primary material product of the big bang was concentrated as exquisitely intense energy. There are many types of energy, but the form most manifest microseconds after the creation was electromagnetic radiation—in simplistic terms, something akin to superpowerful light beams. Then, within the first few moments following the creation, as the universe raced outward, stretching space, a transition took place (a transition the basis for which was discovered by Albert Einstein and codified in that famous equation $E = mc^2$) as energy condensed into the form of matter. A minute fraction of those light beams of energy metamorphosed and became the lightest of elements, primarily the gases hydrogen and helium. Over eons of time, mutual gravitational forces pulled those primordial gases into galaxies of stars. The immense pressures within the stellar cores crushed the nuclei of hydrogen together, fusing them to form heavier elements and, in doing so, releasing the vast amounts of energy we see as starlight. These forces of fusion coupled with those of stellar explosions, supernovae, yielded the ninety-two elements that eventually on planet earth would form building blocks of beings that became alive and sentient. All this was made from the lightlike energy of the creation. Now that is a cause for wonder.

Light beams became alive, and became not only alive, but self-aware, and acquired the ability to wonder. The wonder is not whether this genesis took six days or fourteen billion years or even eternity. The wonder is that it happened. Of that fact there is no debate in science. According to our best understanding of the universe and equally according to the most ancient commentaries on the book of Genesis, there was only one physical creation. Science refers to it as the big bang. The Bible calls it the creation of the heavens and the earth. Every physical object in this vast universe, including our human bodies, is built of the light of creation.

To elucidate the awesome and humbling implications of this incredible transition of light into life, consider the following better understood transition. In one hand I hold a clear glass jar containing the gas oxygen. In my other hand I hold a jar of hydrogen gas. I study the chemistry of these two gases and discover that, under the correct conditions, they can combine to make water, H_2O. Water neither looks nor acts like oxygen and hydrogen, but it is made up of them. When we drink water, we are drinking hydrogen and oxygen in a very special combination. In parallel, we humans and all the matter we see about us may not look like the condensed energy of the big-bang creation, but we are. Einstein's famous equation does not mean that the energy disappears and matter takes its place. No, not at all. What that equation states is that energy can change form and take on the characteristics of matter, just as the hydrogen and oxygen remain hydrogen and oxygen even as they change form when they combine to form water. We are made of the light creation, and no scientist will argue against this. It's not New Age talk or wishful thinking. It's established scientific reality. We, our bodies, were part of the creation.

Our cosmic genesis began billions of years ago in our perspective of time, first as beams of energy, then as the heavier elements fashioned within stars and supernovae from the primordial hydrogen and helium, next as stardust remnants expelled in the bursts of supernovae, and finally reaching home as rocks and water and a few simple molecules that became alive on the once molten earth. We were not observers to this fantastic flow toward life. We were part of it. And unlike the formerly accepted catechism that billions of years passed between the formation of the earth and the origin of life on earth, billions of years during which random reactions in fertile pools of water brimming with energy were theorized to have allowed life to evolve, the discoveries in the 1970s by Elso Barghoorn of Harvard University demonstrated that life began as early as can be geologically recorded. The oldest rocks that can bear fossils, that is sedimentary rocks, already have fossils of microbes, some caught in the act of mitosis, cell division. By the time that the earliest layers of sedimentary rock appeared on earth, nature had already invented life with its ability to survive and reproduce, to store and decipher information. DNA, with its potential to condense a vast molecular library of information within microns of space, was in place and operating. This extraordinary feat of invention and fabrication is recoded in those ancient sediments.

On the primordial, prebiotic earth, there were likely vast numbers of molecules forming and disintegrating. One of them succeeded in climbing the ladder of complexity and became alive. And most wondrous of all, tucked within that fecund molecule that eventually led to the first life, following a myriad of unimagined mutations, was the ability to reproduce. Not only to reproduce, but to do so with some variations in structure. Identical reproduction, a "copying machine," yields stasis. What was needed and what nature produced was a molecule that could reproduce and change, somehow borrowing resources from its immediate environment, until it became a cell. But reproduction is purpose driven, the continuation of the line. That prebiotic molecule, whether by design or by dumb luck, had purpose within it from its inception.

Logically, the first compound that would eventually lead to the earliest life must have had the ability to reproduce. If it did not, then as its molecular machinery degraded, it would have disintegrated. Any beneficial mutations that might have accumulated during its span of existence would have been lost and the trek toward cellular life would then have had to begin again, de novo. Life appeared with purpose already as part of its birthright. This simple undisputed fact is extraordinary.

Even the so-called simpler forms of life, such as microbes, are overwhelmingly complex. The mechanisms of cellular function when studied in detail are not only mind-boggling. They are essentially identical in all forms of life, whether animal, plant, bacterial, or fungal. The likelihood that this complexity could have been chanced upon even once is vanishingly small. Having it arise independently twice by chance is essentially an impossibility. All life must have had a single common origin. But what was that origin?

Could that miraculous flow from inanimate matter to the incredible intricacy of life have been the result of purely random events? Is the incredible not necessarily the impossible?

One answer to the origin-of-life puzzle was spontaneous generation. After all, meat left out, after a very short time, was crawling with maggots. Clearly that was spontaneous generation—the meat gave rise to life. Such was the general opinion until 1860, when Louis Pasteur in a brilliant set of experiments demonstrated that it was flies' eggs, not the meat, that produced the worms. Pasteur had laid to rest the idea of life's spontaneous generation. But Pasteur's discovery only made the puzzle more puzzling. If not spontaneous generation, then what was the source of life? How did the prebiotic inorganic matter become alive?

In 1828, the German chemist Friedrich Wöhler produced what in a sense is considered to be an organic molecule from inorganic substrates according to the laws of chemistry and physics. Inducing a reaction of cyanic acid with ammonia, two inorganic compounds, Wöhler synthesized urea, a substance that until then was associated only with living organisms. Wöhler had demonstrated that the inorganic could undeniably become organic. Of course, urea, though organic, is remote from being alive.

Then in 1953, Stanley Miller, a graduate student at the University of Chicago, produced amino acids in a series of reactions that, like those of Wöhler, started with inorganic compounds. In Miller's experiment, the starting mixture contained those compounds that were assumed to be in the atmosphere and hydrosphere of the prebiotic earth.[4] The discovery by Miller that unguided reactions could yield amino acids was nothing less than sensational. Amino acids are the building blocks of proteins, and proteins are the basic structures of life. Stanley Miller had discovered the key to the origin of life, pure and simple random reactions. Unfortunately, the hoopla was premature. Miller's assumptions that the compounds he used were abundant on the prebiotic earth approximately four billion years ago turned out to be largely false. Furthermore, the reaction and the multitude of experimental variations on that theme never proceeded beyond producing a very few amino acids. The experiment is today considered to be irrelevant with regard to life's origins.[5]

So from where did our primordial ancestors arise? Theories of Darwinian or neo-Darwinian evolution all begin at the stage in which self-replicating organisms are present and abundant. How to get to those bits of replicating life was and is an enigma.

With the discovery in the 1990s that a compound fundamental to all genetic codes, RNA (ribonucleic acid), could perform biochemical functions previously assumed only privy to far more complex proteins (enzymes), the concept of an "RNA world" materialized. RNA, like DNA (deoxyribonucleic acid), is a chain of nucleotides. DNA is double-stranded; most forms of RNA are single-stranded. The presence of hydroxyl groups in RNA makes it prone to hydrolysis and hence far less stable than DNA. This instability notwithstanding, the theory is that before there was DNA there was RNA, and before RNA there were inorganic compounds that via random reactions combined to achieve self-replication with moderate mutations. Then generation followed generation and finally, behold, we have the RNA. There are no data that substantiate this evolutionary theory or how it would evolve into a "DNA world" of which we are a part and to which we owe our existence. Not only is RNA unstable, but also the components required to fabricate RNA are themselves not chemically stable. There is no clue as to how these unstable substrates could have survived and combined to produce the much sought after RNA.

As Robley D. Evans, my M.I.T. Ph.D. adviser and professor of physics par excellence, repeatedly urged, always repeat in summary what you have just espoused. Consider the string of assumptions for which supporting data, if any, are vanishingly scant in an unguided world:

1. A prebiotic atmosphere and hydrosphere existed that could support the reactions among methane, ammonia, carbon dioxide, a few amino acids, and water leading to the complex substrates of RNA. Current understanding

of the prebiotic atmosphere appears to make it hostile to such reactions.

2. The assumed substrates, though diverse in properties and chemically unstable, assembled locally so that they could interact.
3. These substrates combined to form chains of polynucleotides.
4. These polynucleotides became self-replicating molecules able to cull from the adjacent medium (perhaps a slime-like inorganic soup gradually drying in an isolated puddle) the necessary components to rebuild themselves, though of course with slight variations—mutations—that allowed evolution to progress from prebiotic to life.
5. Finally a cell appeared complete with gated membrane to regulate entrance and egress, housing DNA that codes via its four nucleotides for the RNA found much earlier in this process.

Each of these stages presents chemical and physical hurdles for which there are no logical solutions. And yet we have life.

It is time to lay to rest the misguided but popularly believed untruth that in our world, gradual, step-by-step random mutations could have climbed the mountain of improbability and produced the magnificent abundance of the earth's biosphere. To accomplish this goal requires a modicum of elementary arithmetic, some basic high-school-level biology, and a touch of astronomy. But it is worth the effort to bury once and for all the ill-conceived but often unquestioned assumption that random mutations produced life or anything even tenuously related to life.

Stephen Hawking, in his *A Brief History of Time*, the most widely sold science book ever written, teaches the world about the potential power of random events to produce meaningful complex order, such as is found in a work of literature. "It is a bit like the well-known hordes of monkeys hammering away on typewriters. Most of what they write will be garbage, but very occasionally by pure chance they will type out one of Shakespeare's sonnets."[6] It is a compelling premise, but totally off base at least within our universe, and it is life in our universe with which we are concerned. I am surprised that Professor Hawking would have let this slip occur. Nonetheless, it convinced one of the world's leading literary magazines, *The New Yorker*, to devote its Christmas and New Year's cover of 2002 to showing monkeys hammering away on typewriters. As Hawking predicted, most failed to get the sonnet. But, behold, there in the lower right-hand corner is a very happy monkey. He got the sonnet.

I don't know many sonnets. In fact, when I thought about this, I only knew the opening line of one, "Shall I compare thee to a summer's day." There are not quite five hundred letters in that sonnet. All Shakespeare's sonnets are about the same length, all by definition fourteen lines long. Can we get a sonnet by chance? If Hawking says so, it must be true.

But is it? Let's consider 500 grab bags each holding the 26 letters of the English alphabet. I reach into the bag blindfolded and pull out a letter. The likelihood that it will be *s* for the first letter of the sonnet is one chance in 26. The likelihood that in the initial two draws from the first two bags I will get an *s* and then an *h* is one chance in 26 times 26. And so on for the 500 letters. Neglecting spaces between the words, the chance of getting entire sonnet by chance is 26 multiplied by itself 500 times. That seems as if it may be a fairly big number. And it is. Surprisingly so. That number comes out to be a one with 700 zeroes after it. In conventional math terms, it is 10^{700}, or 10 to the exponent power of 700. To give a sense of scale for reference, the known universe, including all forms of matter and energy, weighs on the order of 10^{56} grams; the number of basic particles (protons, neutrons, electrons, muons) in the known universe is 10^{80}; the age of the universe from our perspective of time, 10^{18} seconds. Convert all the universe into microcomputers each weighing a billionth of a gram and run each of those computers billions of times a second nonstop from the beginning of time, and we still will need greater than 10^{500} universes, or that much more time for even a remote probability of getting a sonnet, any meaningful sonnet. Chance does not produce intelligible text and certainly not a sonnet, not in our universe.

But so convincing is Hawking's argument that the students at Plymouth University in Britain convinced the National Arts Council to put up £2,000 (about $4,000 U.S.) to try the monkeys' typing skill. With that stipend they rented a monkey house at the Paignton Zoo in Devon and placed a computer keyboard inside. The *Times* (May 9, 2003) reported on the results under the headline, "Much Ado, but Monkeys Fail Shakespeare Test." For a month, six monkeys hammered away on the keyboard. They failed to produce a single English word. Surprised, since the shortest word in the English language is one letter long? Surely the monkeys must have hit an *a* or an *I* in all their efforts. But think about it. To make the word *a,* a space on each side of the letter is required. That means typing: space, *a,* space. If there are about 100 keys on the computer keyboard, neglecting the fact that the space bar is somewhat larger than the letter keys, the probability of typing space, *a,* space is one chance in a 100 times 100 times 100, which comes out to be one chance in a million. Random guessing in a spelling bee is always a losing proposition. And that is for a single-letter word.

So why does the monkey premise make the cover of one of the world's leading intellectual publications? The reason is distressingly simple. If you are fed from your earliest days the saga that unguided random reactions produced life, then arguing from the major to the minor, certainly you'll believe the untruth that sonnets will come popping out of your random letter generator.

A proverb that *is* actually true and worthy of repeating states: the song a sparrow learns in its youth is its song for life. And we humans, at our deepest emotional level, are not so very different. What we learn in our youth is with us for life. And we all learned that Darwin got it all right, notwithstanding that the article "Did Darwin Get It All Right?" in the world's premier peer-reviewed science journal, *Science,* maintains in the subtitle that unfortunately he did not.[7]

And yet here we are on a beautiful earth brimming with life. From the scorching, more than 100-degree C waters of thermal vents to the frigidity of the Antarctic ice, from sun-soaked Saharan deserts to the blackness of the abyssal oceanic depths, life has staked out its habitat. Life is hearty. It has proven itself to be so. But by random reactions it did not start.

There is a way that the monkeys might produce the grandeur of a sonnet and a random nature yield the wonder of life. But it takes a leap of faith with only the vaguest of foundations upon which to base that leap. And that is the thought that the visible universe is only one of a multitude of universes. In a vast number of universes, say 10^{500}, each trying to produce life, one might have succeeded. In that huge number of universes, we would be in the one that succeeded, while the others would likely be lifeless. Either our universe is a tiny domain, a blip, within an infinitely large universe, or there are a vast number, perhaps an infinite number (if one can combine "infinite" and "number" in the same context), of totally separate universes. With either scenario, each domain or universe has its own set of natural laws. Some sets are similar to our laws of nature, some are totally, even incomprehensibly, different, so different that concepts of time, space, and matter do not even apply. In their place, those unknown universes would be built of dimensions such as igildy, oguldy, uguldy, words that to us are totally meaningless for dimensions we cannot even imagine. Nor could inhabitants of those alien universes, if there are inhabitants, imagine concepts such time, space, and matter.

Sounds far-fetched, but not so. In a valiant attempt to deny the possibility that there might be in our universe any hint of deliberate design or teleology, the May 2003 issue of *Scientific American* devoted its cover to preaching that "Infinite Earths in Parallel Universes Really Exist." No questions asked. They really exist. Several options are presented for how these parallel domains might arise. And the evidence is quite convincing—at least to the author and the journal's editors.

And here is the piece of truth they bring to prove their point. And the following is not buried in the twelve pages of text devoted to this treatise. Rather, so persuasive is the logic that it is highlighted with extensive graphics. It begins in bold red type on a black background and then switches to white type on a black background:

> Evidence: COSMOLOGISTS INFER the presence of Level Two parallel universes by scrutinizing the properties of our universe. These properties, including the strength of the forces of nature and the number of observable space and time dimensions [three spatial dimensions and one time dimension] were established by random processes during the birth of our universe. Yet they have exactly the values that sustain life. That suggests the existence of other universes with other values.[8]

As a scientist, I am embarrassed that such logic can make its way into a widely read scientific journal. Note the convoluted line of reasoning. Our universe has laws of nature made for life. The physical constants that regulate the laws of nature and the behavior of matter are perfect for sustaining life. Such perfection of such a wide variety of laws is vastly unlikely to have arisen by chance on a single throw of the cosmic dice. Therefore there must be a nearly infinite number of universes, each with its own unique set of laws of "nature." Most universes failed to sustain life. Their own individual unique laws of nature are incompatible with complex sentient life. We're the universe that succeeded. All the other options, many lifeless, exist "somewhere" out "there."

Of course the perfection of nature's laws for sustaining life in our universe in no way suggests the existence of other less life-friendly universes. The perfection of our universe's laws of nature is a fact. What it implies is that this perfection is so highly improbable that some explanation other than a simple onetime random event is required. If we are the only universe, then *Scientific American* has inferred that we are indeed living in a designer universe. The Bible claims that the explanation for this perfection is the will of God.

The backpedaling that slips so subtly into the article is instructive. On the journal cover we are told that these "infinite earths in parallel universes really exist." In the text we learn that perhaps they exist. Their being is "inferred."

Of course there is a simpler option than a nearly infinite array of sterile universes. That option is that we are the only universe and we did not arise by chance. We were designed. But that of course is not acceptable logic in a materialist view of existence. However the author of the less than profound logic upon which that scientific article is based has something good going for him. And that is you can never see out of the universe within which you reside. You can never directly contact the other universes, so you can never ever check out the proof of your theory. It's a massive leap of faith that makes the Bible's scenario seem timidly conservative by comparison. We will look at this more closely in the next chapter.

Either way, Bible or multiuniverses, at the end of the day we are still mired in the ultimate of conundrums: Why is there a universe, or even a multitude of universes, to host beings who ask these questions? Why is there something rather than nothing? Why does existence exist?

Statistics reveal the numerical paucity for randomness being the source of the stable order evidenced in life. The Torah brings the same information in the subtle wording of "And there was evening and there was morning," the closing phrase of each of the six days of the first week in Genesis. But the sun is mentioned only on day four (though the ancient commentaries tell us that the sun was there earlier, but only became visible in the sky on the fourth day). Evening means sunset. Morning means sunrise. No sun, no sunset and no sunrise. So how can the Bible justify the statement that "there was evening and there was morning" on those days prior to the sun's mention on day four?

Almost a millennium ago, the biblical commentator Nahmanides realized that there must be a deeper meaning to the words "evening" and "morning." He taught that the root or implied meaning of the Hebrew word usually translated "evening," *erev,* is "mixture," "chaos." As the sun sets, he reasoned, vision becomes blurry, mixed. The implied meaning of the word usually translated "morning," *boker,*[9] is just the opposite. As the sun rises, vision becomes clear, orderly. Individual objects and colors can be discerned. The implied meaning of *boker* has within it the concept of order. The flow in simple terms is from P.M. to A.M. But the deeper meaning, the far more significant truth, is that six times over, at the conclusion of each day of creation, there was a remarkable flow contrary to what is normally observed in an unguided nature. Normally, in all events, order degrades to disorder. That is why leaves decay on the ground and a cup of hot tea becomes cool as it remains on a table. But in this particular part of the universe, the opposite occurred, and the Torah emphasizes this six times over in the subtle language "And there was evening and there was morning." The ordered complexity of life arose from a mix of rocks and water and a few simple molecules, and even more remotely, from the chaotic burst of energy that marked the big-bang creation, an energy brimming with potential, only awaiting the organizing realization at the word of God: "And God said, let there be . . ." In this part of the universe chaos gave way to life.

But what about life's origins? A universe with laws of nature excellent for sustaining life may not have laws of nature amenable to the inception of life. The starting of life likely has vastly different physical and chemical requirements than those needed to sustain that life following its inauguration. Can random mutations in our universe actually have produced the ordered complexity of life or even a viable protein that is so well designed for sustaining life?

But let's be more conservative in our quest and accept that somehow life started and now we need that early form of life to mutate and climb step by step the fabled mountain of improbability. Mutations that are to be passed on to the next generation must occur in the genetic material, that is, in the DNA of the reproductive line. Such a mutation might result in a variant (mutated) protein that might produce a new effective organ, say, a system leading to a kidney or the precursor of a pump that might develop into a heart. The neo-Darwinian concept of evolution claims the development of life resulted from random mutations in the DNA that yielded these varied organic structures. Some of the variations were beneficial, some not. The rigors of the environment selected for the beneficial changes and eliminated those that were detrimental.

It's a persuasively devised theory, but let's look at that process rigorously, especially with the insights of molecular biology. The building blocks of all life are proteins. And proteins are precisely organized strings of amino acids. Information held in the DNA determines which and in what order the amino acids are formed to yield the end product, the protein. If the DNA mutates, we get a different amino acid and hence a different protein. And now comes the problem of random mutations in the neo-Darwinian theory of evolution.

The genetic system of all life is totally coded. An example of a code would be the Morse code sounds "dot dot dot dash," which look, sound, or seem nothing like the letter *v* for which they are code. If you didn't know that the sequence of sounds "dot dot dot dash" represents a *v*, you wouldn't have even a hint as to its meaning. That is one purpose of a code. And so it is with the information encoded on the DNA chromosomes. The data on the strands of DNA (the chromosomes) in our cells contain that crucial amino acid and protein–building information as assorted groupings of four different nucleic acids. Nucleic acids have absolutely zero physical resemblance to either amino acids or proteins. The information is totally coded.

In nature, this lack of similarity between code and final product ensures that there is no logical feedback from protein or amino acid to DNA. Information flow is one-way: DNA to amino acid to protein. New mutant variations of proteins arise through mutations (changes) in the sequencing order of the nucleic acids on the DNA with no physical hint of the final protein product. These random, unguided mutations are the determining factors in gain or loss of that next generation.

In all known life, there are primarily twenty different amino acids. Stringing these twenty amino acids together in varied sequences produces varied proteins, just as intelligently stringing together the twenty-six letters of the English alphabet in varied sequences will produce varied sentences and sonnets. Scientific literature suggests that all of life is made from varied combinations of several hundred thousand proteins. Humans have in the order of eighty thousand proteins. (The estimated number of proteins in humans varies among laboratories reporting their results.) Other forms of life have different numbers of proteins. But all life, whether animal, vegetable, microbial, or fungal, draws from the same "bag" of functional proteins. That being the case, it is not surprising that we humans contain some of the same proteins found in plants and animals that are very different from us. Proteins, other than those within the cluster of those used by viable life, form by mutations on the DNA sequencing of nucleic acids. Cells actually have a highly sophisticated mechanism that checks for mutations early in the molecular progression that leads to protein formation. Upon discovery of a mutation, the molecule is either sent back for revamping or destroyed. But some mutations slip by the checkpoint. These may be either useful, neutral (adding no selective advantage for survival), or lethal. A painful example of a mutation leading to a lethal protein would be a mutation that becomes a precursor for cancer.

So we have these few hundred thousand proteins that are viable in life. Others appear not to be. But let's say we are off in our estimate. In place of a few hundred thousand viable proteins, let there be 100 million or a billion or even a trillion viable proteins. And now to the crucial numbers.

An excellent review article published in the *Proceedings of the National Academy of Science* compared the rates of point mutations in the genomes of forty mammal pairs with separations ranging from 5.5 to 100 million years ago.[10] The observed rates of mutation were quite similar among all the animal pairs, averaging at 2.2×10^{-9} mutations per base pair per year. This rate relates to point mutations, that is, alterations in the sequencing of the individual nucleic acids along the chromosomes of the DNA and not to sequence repetitions in which long strings of nucleic acids are replicated on the DNA. At this rate, the difference between humans and our closest genetic relative, the chimpanzee, would be approximately 30 million nucleic acids (base pair) differences, a number very similar to what is biologically measured.

Let me be transparent. I am not debating how a fin could mutate and eventually become a foot. Fins and feet have many structural elements, especially bones, in common. With a stretch of imagination, we can envision a series of changes, such as sequence repetitions, that would morph a fin into a foot. But how do random mutations initially produce the genetic information that would lead to the molecular structure of any sort of bone? Or muscles that eventually become the pumps that are prelude to a heart?

Strings of proteins vary in length from a few hundred to a few thousand amino acids. Consider a relatively short protein, such as one 200 amino acids long. Into each of the 200 spaces along the protein any one of the 20 amino acids found in life can fall. That means the total number of possible combinations is 20 times 20 times 20 repeated 200 times. The result is 20 to the power of 200, or ten to the power of 260 (10^{260}), a one with 260 zeros after it, or a billion billion billion repeated 29 times. From this vast biological grab bag of options, we are told that nature, by random chance mutations, has been able to form the few hundred thousand proteins useful to earthly life and upon which nature could exert its selective pressures.

Let us assume that the entire hydrosphere, all of the approximately 1.4×10^{21} liters of water in all the oceans and icebergs and lakes on earth, was imbibed in biological cells each weighing a billionth of a gram. We would have had 10^{33} cells reproducing, mutating, actively moving this grand process of evolution. If each cell divided each and every second since the appearance of liquid water on earth some four billion years ago, the total number of mutations, or stated another way, the number of evolutionary trials, would be 10^{50}. Although vast, this number pales when compared with the 10^{260} potential failing options for a single protein. Hitting upon the useful combinations did not, and could not, and will not happen by chance.

All biologists enamored with neo-Darwinian evolution know this truth. Their hopeful reply goes along the line that, although we now have a DNA world, other worlds may have been possible, and DNA, being the first to form and survive, merely took over. Other systems might have used other types of proteins that we see as lethal or useless in today's DNA world. There is no evidence that this is true; however, let us assume its truth. Now we have the DNA-dominated world we know. And so we are back to the above calculations as the first form of life, a microbe, mutates and either advances or perishes as it starts to climb the mountain of improbability by random mutations on the DNA that in time will lead to kidneys, bones, liver, heart, eyes, brains, mind, sentience. It has to choose randomly from the vast hyperspace of possible biological combinations the tiny fraction that are beneficial or at least neutral. Clearly there must be other factors that limit the types of mutations that can occur. There are, but not as randomly as materialist biologists would have it. And that is the entire point. Nature is skewed toward life.

And that is exactly what one of the most widely used biology textbooks, *Biochemistry,* by D. Voet and J. Voet, states, though in subtle wording: "Keep in mind that only a small fraction of the myriads of possible peptide sequences are likely to have stable conformation. Evolution has, of course, selected such sequences for use in biological systems."[11] Just how did "evolution" become so clever that it could "of course select" from the "myriads" of failures the few that function?

Jon Seger, professor of biology at the University of Utah, tells us how, of course following the central dogma of Darwinian adherents and neglecting the statistical improbability of its being driven by random mutations:

> Within a population, each individual mutation is extremely rare. . . . But huge numbers of mutations may occur every generation in the species as a whole. [That is because each member of the population may only have a few mutations, but when multiplied by the total number of mating members, the total number of mutations per generation can be very large.] . . . The vast majority of the mutations are harmless or at least tolerable and a very few are actually helpful. These enter the population as exceedingly rare alternative versions of the genes in which they occur. . . . Very small effects on survival and reproduction may significantly affect the long-term rates at which different mutations accumulate in particular genes. They just accumulate where needed, first one, then another and another over many generations. Although getting two or more new cooperating mutations together in the same genome may take time, they will eventually find one another in a sexual species [and since by getting together they provide an advantage over the former configuration, the organism with this new advantage will now flourish relative to the less adapted neighbor], assuming they are not lost from the population.[12]

All that Professor Seger writes is largely true. Indeed "the vast majority of the mutations are harmless or at least tolerable," though many may be lethal. But even if none were lethal, the problem is not the ultimate "natural selection" according to the rigors of the natural environment, selection between good and better, strong and stronger, more fertile and less fertile. Those selections are at the final stage of the process. And we see it verified each time as a strong lion vanquishes or kills a weaker competitor for the right to fertilize the females of the pride. But first nature must produce those variations of advantage via random mutations of the nucleic acids on the genome that change the chain of amino acids that form the protein that alters the viability of the "animal."

The statistically unrealistic possibility that the fabrication of viable proteins could have occurred by unguided random mutations is simply ignored. That life developed from the simple to the complex is, in my opinion, a certainty. What drove that development is the central debate.

Simon Conway Morris is professor of evolutionary paleobiology at Cambridge University, as mentioned, and Fellow of the Royal Society of England. He is arguably the world's leading living paleontologist. In his book *Life's Solutions,* Conway Morris states the conundrum perfectly: "The number of potential 'blind alleys' is so enormous that in principle all the time since the beginning of the universe would be insufficient to find the one in a trillion trillion solutions that actually work. . . . Life is simply too complex to be assembled on any believable time scale. . . . Evolution [has the] uncanny ability to find the shortcuts across the multidimensional hyperspace of biological reality."[13]

Conway Morris, as with most scientists active in the field of developmental biology, accepts the quite firmly established principle that life developed from the simple to the complex. The query with which he deals in his book, and with which we struggle here, is a search for the mechanism behind the flow that has produced the magnificent biosphere of which our bodies are a part.

As such, Conway Morris opens his book with a disclaimer: "If you happen to be a 'creation scientist' (or something of that kind) and have read this far, may I politely suggest that you put this book back on the shelf. It will do you no good. Evolution is true, it happens, it is the way the world is, and we too are one of its products. This does not mean that evolution does not have metaphysical implications; I remain convinced that this is the case."[14] Morris uses the word "evolution" in the European sense, allowing for the "metaphysical" to have been active in the orchestration of the process. Evolution in the American sense, however, insists that the mechanism that drove life's development was totally random mutations on the genome that were then selected for or against by the ability of the mutant organism to survive the trials of nature. In American-ese evolution and "metaphysical implications" are mutually exclusive. Conway Morris is by far not the only accomplished scientist who realizes that there is plan inherent in the development of life.

The late George Wald, Noble laureate and professor of biology at Harvard University, may have provided us with the answer to the wonder of life in an essay he wrote titled "Life and Mind in the Universe" for the 1984 Quantum Biology Symposium:

> It has occurred to me lately—I must confess with some shock at first to my scientific sensibilities—that both questions [the origin of consciousness in humans and of life from nonliving matter] might be brought into some degree of congruence. This is with the assumption that mind, rather than emerging as a late outgrowth in the evolution of life, has existed always as the matrix, the source and condition of

> physical reality—the stuff of which physical reality is composed is mind-stuff. It is mind that has composed a physical universe that breeds life and so eventually evolves creatures that know and create: science-, art-, and technology-making animals. In them the universe begins to know itself.[15]

This almost mystical analysis of life is from the same George Wald who thirty years earlier in an article in *Scientific American* declared with no equivocation that life is indebted totally to pure random chance for its existence: "The important point is that since the origin of life belongs in the category of at-least-once phenomena, time is on its side. However improbable we regard this event, or any of the steps which it involves, given enough time it will almost certainly happen. . . . Time is in fact the hero of the plot."[16] Interestingly, twenty-five years after its publication, *Scientific American* retracted that article unequivocally stating: "Although stimulating, this article probably represents one of the very few times in his professional life when Wald has been wrong." The retraction stated "that merely to create a single bacterium would require more time than the universe might ever see if chance combinations of its molecules were the only driving force."[17]

Macromolecules have been found to possess the amazing ability to "self-assemble." This ability is built into the structure of the universe. Wald's epiphany occurred when, in conducting the research by which he earned the Nobel Prize, he elucidated a portion of the mind-boggling complexity in the series of reactions at the eye's retina that allows the picture in the mind to remake itself ten or fifteen times a second. At the quantum level it appeared that mind, intelligence, was somehow embedded in the process.

Equally brilliant Steven Weinberg did not have a similar epiphany when delving into the physics that might lie behind the creation of a universe suitable for life or complexity in any form. Creation physics always leaves the opportunity for diverse explanations of beginnings: quantum fluctuations, multiple universes, an eternal megauniverse within which we are but a blip as a microuniverse. And even then "at the end of the day," atheist Weinberg acknowledges that the ultimate question remains. Why is there anything rather than nothing?

This is science, not theology, speaking. But it is also theology. "In the beginning was the *logos* [*logos* is the Greek word for 'logic,' 'intellect,' 'word']" (John 1:1). A few hundred years before John, "With the word of God the heavens were made" (Ps. 33:6). And a few hundred years before that, we find the opening sentence of the Bible, Genesis 1:1. The traditional translation of that crucial verse is "In the beginning God created the heavens and the earth." That is the traditional reading, but as Frank Lloyd Wright so pithily pointed out in his seminal book *The Natural House,* tradition can be an enemy. A more accurate rendering of the verse is "With a first cause (*B'raisheet*) God created the heavens and the earth." "In the beginning" is the mistranslation of the first word of the Bible, the Hebrew word *B'raisheet.* The error was introduced by the Septuagint twenty-two hundred years ago and then carried through to the Latin Vulgate and finally the King James and other English translations.

B'raisheet in its simple sense translates as "In the beginning of." But there is no object in the Hebrew text for the preposition "of." It would read, "In the beginning of God created the heavens and the earth." In the beginning of what? So the Greek and the Latin merely deleted the "of," which of course is ridiculous and borders on the heretical, as if these ancient translators felt they could better state the facts than did the Bible. The compound nature of *B'raisheet* holds the clue. Several years ago a skeptical student challenged me on this, claiming that if indeed the *B* in the word *B'raisheet* truly stands alone, then the text should indicate this. He nearly burst into tears when I showed him a Torah scroll. Every (every) Torah scroll has the *B* of *B'raisheet* written twice the size of the remaining letters of the word. (This truth precedes me by twenty-five hundred years in Proverbs and was further elucidated by the preeminent theologian Rashi a thousand years ago.)

And what was that first cause? The twenty-one-hundred-year-old Jerusalem translation of the Bible into Aramaic, a sister language of Hebrew, holds the answer. *B'raisheet* is a compound word meaning "with" or "using" (*B'*) "a first cause" (*raisheet*); hence, "With a first cause [of wisdom] God created the heavens and the earth." That the "first cause" is defined as wisdom arises from Proverbs (8:12, 22–24): "I am wisdom. . . . God acquired me [wisdom] as the beginning of His way, the first of His works of old. I [wisdom] was established from everlasting, from the beginning, from before there ever was an earth. When there were no depths I [wisdom] was brought forth." Wisdom, the totally metaphysical emanation from the Creator, yielded the big-bang creation of the physical universe within which we dwell.

With wisdom (Proverbs) and logic (John) and mind (Wald) or, in the language of quantum mechanics, information (J. A. Wheeler) as the essence of existence, the puzzle of the origin of sentient life able to be aware of the wonder of its own existence is solved. Wisdom is ubiquitous, the substrate of every particle of the world and most evident in the brains and minds of humans as we puzzle over our cosmic origins. The success of life is indeed "written into the fabric of the universe."

Our Cosmic Genesis in a NutShell

God → wisdom → big-bang creation of energy → matter → life → brain → mind and sentience

Concisely stated, the wisdom of God embedded in the energy of the big-bang creation laid the basis for that seemingly inert energy to metamorphose and become alive. And not merely alive, but even more than that—to become alive and brimming with the sentient awareness of being alive. As Professor Wald stated so well: "It is mind that has composed a physical universe that breeds life and so eventually evolves creatures that know and create: science-, art-, and technology-making animals. In them the universe begins to know itself."

Within every piece and aspect of the world, there lurks at its foundation the essence of wisdom, or mind, an emanation of the Force that brought it into being. As bizarre as it may seem, we will discover that the world in a very real sense has a "mind of its own"!

To understand how that dynamic Force manifests Itself in the ever changing world It created, we turn to the only two sources of relevant information: nature, that is, the world around us, and the Bible. Both provide a confirmation that God's essence is as vibrant as the world itself. In this sense the study of nature is as much a study of God as is the study of the Bible. The eighteenth-century theologian known as the Scholar, or Gaon, of Vilna taught that when the Torah was given on Sinai, it split into portions. Only one portion was retained as the written words of the Bible. The other portion was hidden in nature. And only when we finally discover that part of Torah that was sequestered in nature will we be able to fully understand the word of God.

Psalm 19:1 declares: "The heavens proclaim the glory of God; the sky declares His handiwork." As a person who has worked in the sciences most of my life, in both physics and the earth and life sciences, I could not agree more fully. The hand of God reveals itself in the grandeur of galactic space as well as in the details of an atom. In both realms, though they differ vastly in dimension, we can learn how God acts within this world It created. As nature provides the background for the biblical writings, so the earth provides the substrate for the miracle of life. In the next chapter we will study a few aspects of the earth that make it such a friendly place for life.

THREE

The Unlikely Planet Earth

Finding a Friendly Home in a Challenging Universe

The greatest evidence for a Creator is the uniqueness of the creation itself. We must put this quality into proper perspective. Just as it was essential to get past the ill-formed idea that pure randomness can climb the mountain of improbability (and so produce by pure random trials a sonnet or a microbe), it is equally essential to lay to rest the supposition that our planet is merely an ordinary, "run of the mill" planet, one among a multitude of earthlike planets in a universe brimming with life.

Innumerable journals and books, secular and religious, my previous books included, marvel at how well the laws of nature are tuned to produce a nature that is amenable to life. If the strong nuclear force were a bit more or less forceful, if electromagnetic forces varied by a few percent, if there were four or two spatial dimensions instead of the three within which we reside (length, width, height), the stability of the physical world would be gone. Without the Pauli exclusion principle that teaches that no two particles can occupy identical positions at the same time, and the quantized electron orbits, stable atoms could not form. There'd be no predictable chemistry or chemical reactions, no life. Nuclear chaos would be the order of the day.

But our universe has just the right laws of nature, the just right physics and chemistry to sustain life. Perfection so dramatic and extreme has caused, as noted previously, the secular speculation that there must be vast numbers of universes with different laws of nature, a few fit, most not fit, for sustaining life. Our just-right universe is the rare random success story for life among the many random failures. No designer required. Only a vast number of universes and random chance stumbling eventually on the laws of nature needed for complex life.

There is, however, an additional level of just-rightness required for a biosphere to flourish. And that is that from those perfectly designed laws of nature there evolved an environment having a climate amenable to the nurturing of life. Needed is a stellar system with a planet at just the right distance from just the right star, with just the right mass. The solar system and the earth are it.

Did random chance demand that there be one or even many well-suited stellar systems in our universe? The answer is not at all obvious. Recall the one case out of 10^{700} for getting a Shakespearean sonnet by chance. Compare the magnitude of that number with the number of stars in the visible universe: approximately 10^{22}. Now, 10^{22} is a huge number, but it pales when contrasted with 10^{700}. Obviously getting sonnets and evolving life are two very different concepts. But both are measures of chance producing a desired arrangement. With the sonnet-writing monkey trials, there are 26 variables (the letters of the alphabet) that had to align in 500 distinctive ways. In our universe to win the lottery for life there are 10^{22} options, the 10^{22} stars in the visible universe, each offering its own unique environment. Around how many of those stars will there be planets with conditions favorable to life? If there are, for example, 15 parameters essential for producing a planet or moon with a life-nurturing climate, and only 3 or 4 percent of the total cosmic range of each of those 15 variables places it within a domain fit for life, then the likelihood of life arising by chance in our universe becomes marginal.

In the 1950s, Frank Drake, an astronomer, put together an equation having seven variables that, when multiplied together, gave an estimate for how many planets with advanced life (on the level of intelligent life) might exist in our galaxy. Among the variables in Drake's equation are two factors estimating the fraction of all potentially habitable planets that are actually inhabited by intelligent beings. Both of these factors are wildly speculative and embody an a priori assumption that life is actually distributed in the universe. The latter is a totally unwarranted postulate. As Nobel laureate Enrico Fermi is quoted as wondering, "If there really is life out there, where are they?" Fermi was alluding to the probability that if there is indeed life brimming among the one hundred billion stars of our galaxy, the Milky Way, then based on the time taken for complex life to have developed on the earth, statistically if other earthlike planets exist in the Milky Way, intelligent life should already have developed on them. By now, we should have found some indication of their existence. And we haven't.

A calculation similar to Drake's, but taking a much more modest (and I think a more realistic) approach, is to estimate how many planets in the entire universe might have conditions amenable to advanced, animal-level, carbon-based life. We know that at least one has. We call it the earth. With this approach we avoid having to guess at the likelihood of life starting elsewhere or anywhere. I speak of our concept of complex life as being carbon-based, because carbon is the only one of the ninety-two naturally occurring elements able to form the variety of long-chained, complexly configured molecules associated with life. The environmental conditions required to induce life to start, circumstances that could cause a prebiotic world composed of inanimate rocks, water, and few simple molecules to metamorphose and take on the characteristics associated with the information-packed complexity of life, are a total mystery. Since we have no direct knowledge as to how life started on earth, let's just explore the characteristics necessary to create an environment amenable to sustaining advanced life, once life has already invented itself.

What might some factors be that would make a domain habitable? Temperature, for a start. The approximate temperature range in our solar system is from 5,500 degrees C on the surface of the sun (15 million degrees C at the center) to minus 270 degrees C at the outer planets. The temperature of space, the minus 270 degrees C, is a mere 3 degrees Kelvin above absolute zero (absolute zero, the temperature at which all molecular motion ceases, is minus 273.2 degrees C, or minus 459.7 degrees F). Life as we know it falls within a temperature range of approximately 100 degrees C, starting at the temperature at which water becomes liquid and ending at its boiling point. That's 100 out of about 5,800, or 1.7 percent. Of the temperature range experienced in the solar system of sun and planets only 1.7 percent is fit for life. Of course, if we take the total temperature range in the solar system, including the sun's center, then the slice suitable for life falls to 100 out of 15 million, or one part out of one hundred thousand. That's a real constraint on positioning. Let's work with the more life-friendly 1.7 percent.

The shape of our galaxy turns out to be highly skewed toward fitness for life. There are spherical, spiral, and elliptical galaxies, star clusters, small and large, and colliding galaxies. Spiral fits the needs. The others are marginal at best, as we shall see. Then there's the parent star's location within the galaxy: near the center or farther out? Is the star embedded within a spiral arm or between the arms? For life we need planets circling the star. How far from the star are they? Close enough for the star's radiant warmth to keep water liquid, but not so close that the water boils away. And which of the ninety-two elements are present? This varies greatly among stars. And then there is concern for the planet's water content, continental drift, tectonics and mountain chains, dry land and oceans. All are critical as life-sustaining factors. Does it have a large moon; one or more giant planets in the outer system, but no giant inner planets; strong enough gravity to retain an oxygen-rich atmosphere but not so strong as to hold lighter toxic gases; a nearly circular planetary orbit; moderately rapid planet rotation; and stability and tilt of the rotational axis? The list goes on, and each aspect plays an essential role in the construction of a region fit for life. The jargon in the trade is the "habitable zone." Creating a habitable zone around a star is not a trivial task.

Copernicus succeeded in conceptually removing the earth from the center of the universe. With that came the idea of many worlds, only one of which contained the sun and earth. It led to the erroneous though popular concept that our solar system is just one of a myriad of stellar systems fit for life. The sun may be a main-sequence star as far as age is concerned, but its positioning within the galaxy, its metal content, and its size relative to other stars are far from average. Modesty aside, we're neither average nor mediocre. And just to prove it, consider the following.

Our galaxy, the Milky Way, is one of approximately a hundred billion galaxies in the visible universe. Its stars form a spiral. Spherical, elliptical, irregular, globular, and colliding galaxies also populate our vast universe. Spiral is what we want. The other configurations are populated with stars that are either mostly too young, and hence have too low a concentration of the elements necessary for life, or too densely packed with adjacent stars, and as a result are too often bombarded with life-threatening radiation from those neighboring stars.

Complex advanced life as we know it, or even how our fantasies might imagine it, requires a stable environment, rich in elements able to join together to form the complex molecules associated with even the simplest of microbes. The immense pressures during the initial moments following the big-bang creation forced hydrogen, the lightest of elements, being composed of a single proton, to fuse with two neutrons and another proton to form element number two, helium. In this transition from hydrogen to helium there is a highly unstable stage as the particles are sequentially joining together. Due to that instability, only a limited amount of helium could form. The result was that during the first few moments following the creation, only the three lightest elements, hydrogen, helium, and a minute amount of lithium, formed from the energy of the big-bang creation.

Heavier elements were only to take shape later as stars reprocessed those earliest nuclei. In *Genesis and the Big Bang,* I discuss in detail this transition from light to heavy elements and the wondrous implications of that hydrogen-to-helium bottleneck. Briefly stated, if the synthesis of the ninety-two elements from the energy of the big bang had been physically simple, without the nuclear block due to the intricacy of the hydrogen-to-helium transition, then there'd be essentially no hydrogen in our universe. It all would have fused into heavier elements, culminating in superstable iron, element number twenty-six. Stars attain their energy from the fusion of hydrogen into helium within the massive pressures of the stellar cores. In that fusion, the mass of the helium formed is less than the hydrogen consumed. The lost mass has reverted to its primordial form of energy as Einstein had predicted in the most famous equation ever written: $E = mc^2$. That fusion is crucial beyond its energy-forming function. It also produces the helium needed for the following stages of heavy-element formation so essential for the development of life. Had there been no hydrogen remaining after the first few moments following the big bang, had all the hydrogen fused into helium in those initial moments, then no hydrogen would have meant no energy source to power stars. No stars means no long-duration, stable sources of energy and no variety of heavier elements, both necessary for the gradual development of advanced intelligent life. Just another aspect of how the universe is tuned for life.

Over the eons of stellar lifetimes, as the massive pressures within the cores of stars fuse hydrogen nuclei to form helium, the internal hydrogen concentrations fall. Once a star has consumed its hydrogen fuel, the massive gravity of the star's gases may cause the star to implode and then in rebound explode as a supernova, spewing its debris into space. This literally is stardust. The colossal forces in the explosion build the heavier elements, like LEGO blocks one onto the other, as they move up the list of the elements. Essentially all the elements of our universe beyond the mass numbers of the three lightest elements, hydrogen (one), helium (two), and the initial smidgen of lithium (three), were formed through amalgamation in the stellar furnaces as stars consumed their hydrogen fuel and then exploded, seeding space with their element-rich dust so essential for life. This metal-rich stardust may then have mixed with clouds of the primordial hydrogen originally formed in the big bang. Once again, through the forces of gravity, the mix of gases and stardust can draw together to form a second- or third-generation star. Such is the origin of our sun and its life-giving traits. The solar system is so very rich in the heavier elements that it cannot be a first-generation star.

When analyzing the relative abundances of the elements in the universe, it is not surprising that hydrogen and helium, the two lightest elements, rank first and second (approximately 75 and 25 percent, respectively). That oxygen and carbon, two life-essential elements, are the third and fourth most abundant elements is surprising. With carbon being the most abundant element in the universe that is solid in the temperature range that water is liquid, one begins to wonder if the universe knew we were coming. The electron configuration of carbon allows it to bond with itself in a variety of ways that exceeds all other elements, a trait necessary for the complex molecules associated with even the simplest forms for life. Yet the building of carbon from helium is far from straightforward, requiring multiples of helium to combine and become stable units within the fraction of a fraction of a second before breaking apart.

Sir Fred Hoyle, the knighted astronomer, a scientist well known for his work in the theory of star and element formation and also a former avid agnostic with a tendency toward atheism, predicted that, since carbon is so abundant, there must be a "resonance," an energy level leading to a carbon isotope, that favors its formation. The resonant energy level was discovered. Hoyle became a believer. In an article in *Engineering and Science,* the quarterly magazine of one of the world's premier scientific universities, the California Institute of Technology, Hoyle wrote:

> Would you not say to yourself, "Some supercalculating intellect must have designed the properties of the carbon atom, otherwise the chance of my finding such an atom through the blind forces of nature would be utterly minuscule." Of course you would. . . . A commonsense interpretation of the facts suggests that a superintellect has monkeyed with physics, as well as with chemistry and biology, and that there are

> no blind forces worth speaking about in nature. The numbers one calculates from the facts seem to me so overwhelming as to put this conclusion almost beyond question.[1]

Such an expression of fervent conviction by an acclaimed scientist and former skeptic, induced by the facts of nature, should give atheists pause to reflect, to reexamine their faith-held atheism. Hoyle's reaction to the facts is not so very different from that of the former atheists' atheist, philosopher Antony Flew, who relinquished his firmly held faith in atheism when confronted with information that plainly contradicted the basis of that ill-founded faith.[2]

The significance of carbon in life cannot be overstated. It accounts for half the dry weight of our bodies. Carbon along with oxygen, nitrogen, calcium, potassium, iron, and a multitude of other life-essential elements are dependent upon star formation and explosive death (supernovae). As such, they are not distributed uniformly throughout our universe and not even uniformly within a given galaxy. Galaxy type and position within a galaxy play a strong determining factor in the local metal content. The stars of elliptical galaxies are often first-generation stars, many being almost as old as the universe. As such, they are poor in heavy elements.

Globular clusters of stars are also typically metal-poor ("metal" is astronomers' terminology for elements heavier than helium, that is, those elements formed within stellar furnaces). In addition, the dense packing of the stars within the cluster creates problems for life. Within a radius of 10 to 15 light-years (one light-year, the distance light travels in one year, is approximately 9.5 trillion kilometers; trillion in the American use, a million million), a cluster might contain in excess of a thousand stars. The same size sphere around our sun contains less than two dozen stars, with the nearest star, Alpha Centauri (actually a system of three stars locked in gravitational orbit) being 4.3 light-years (40 trillion kilometers) distant. The close proximity among stars in the clusters results in devastating local instabilities. Frequent nearby supernovae and other active sources of intense, sterilizing radiation along with stellar gravitational interactions perturbing planetary orbits spell death for nascent life struggling to gain a planetary foothold. Our star, the sun, is located in what might be termed as a stellar desert.

Collisions among galaxies, though not common at least from evidence within the visible universe, can result in what astronomers refer to as supernovae factories, with stellar explosions every few years. The gestation period for the development of complex animal life exceeded three billion years on earth. If this timing is typical, then the nearly continual onslaught of ruinous radiation associated with these stellar explosions would obviate the potential for advanced life.

Spiral galaxies are fit for life because spiral galaxies provide a range of conditions among which are those optimal for nurturing life. The spiral structure offers a variety of stellar concentrations throughout, from high densities near the galactic center and within the spiral arms to very much lower densities of stars in the spaces between the arms. Depending upon how the outer limit of the Milky Way is defined, the diameter of our galaxy is between 80,000 and 100,000 light-years. Our sun lies between two spiral arms, some 27,000 light-years, or approximately two-thirds of the way, out from the Milky Way's center. Relative to the thickness of the Milky Way's galactic disk, the solar system is just above (or below, depending upon one's orientation) the center of the 6,000-light-year-thick central plane.

The earth revolves around the sun at approximately 30 kilometers per second. The sun and the solar system revolve around the center of the Milky Way at some ten times that speed. All revolutions are in the same direction. A deep challenge remains to discover why all components of the universe revolve or rotate. Angular momentum, as the forces of rotation are categorized, is always conserved unless acted upon by an outside force. What force acted to induce the rotations observed is an unsolved puzzle. A universe without rotation would mean a universe without planets and without life. All solids and gases would be gravitationally drawn directly into the parent star rather than revolving around the parent star.

That the sun is located between two spiral arms means that it is in a region of low stellar density, with all the life-supporting features that that implies, most important of which are the fewer destructive interactions originating from neighboring stars. Being far from the galactic center shields us from the radiation-rich star-forming processes of the central region and distances us from the devastatingly powerful gravitational reach of the central black hole.

The edges of most galaxies are metal-poor; our positioning within the galaxy places us in a metal-rich area. These heavier elements are needed not only for the molecular structures of life. They also are essential for the formation of rocky planets such as our earth.

Our sun is about halfway through its theoretical ten-billion-year life expectancy. All indications are that its energy output has remained fairly constant over the past billion years, neither frying nor freezing the earth's surface. More than 90 percent of all stars are smaller than the sun. Smaller generally means longer-lived, more time for "evolution" to work its processes, provided it is not so long-lived that it is a first-generation star. That would make it metal-deficient.

A planet's distance from its parent star has a crucial influence on its suitability to nurture life. Proximity to the parent star results in gravitational drag on the planet's rotation, slowing the rotation through what is termed tidal locking. Gradually the period of rotation becomes similar to the period of the planet's revolution around the star. The outcome is that the same side of the planet continuously faces the star. The moon relative to the earth and Mercury relative to the sun are cases of tidal lock. We always see the same side of the moon. Mercury revolves around the sun in eighty-eight days and rotates on its axis once in fifty-nine days. On Mercury, the extraordinarily long day results in the sun side baking at 400 degrees C while the shaded side drops to minus 170 degrees C. Our twenty-four-hour rotation spreads the sun's heat. Our nearly circular orbit keeps that solar input nearly constant year-round. These are two of the many life-giving characteristics of the earth-sun system.

It is not surprising that Mercury is tidally locked to the sun. At 58 million kilometers from the sun, it is the closest planet to the sun. What is unexpected, on a superficial analysis, is that Venus is also tidally locked to the sun. Venus, at 108 million kilometers from the sun, is twice the distance from the sun compared to Mercury, and still the force of the sun's pull has locked it into rotational sync. Venus is just under 30 percent closer to the sun than earth. Its period of rotation is 243 days, while its revolution around the sun takes 225 days. Aside from Mercury, Venus is the only other solar planet thus locked. Had the earth been closer to the sun, it too would have suffered a similar fate, one side baking and one side freezing.

A glance at the spacing of the seven inner planets tells wonders. (Neptune and Pluto are not considered here, since their orbits cross and therefore interact. From 1979 to 1999 Pluto was actually closer to the sun than was Neptune.) The orbit of each of the seven inner planets places it at approximately twice the distance to the sun as the planet in front of it. For example, Mercury is 58 million kilometers from the sun, and Venus is 108 million kilometers. Mars's distance, 228 million kilometers, is approximately twice the distance of Venus from the sun. The asteroid belt similarly is twice the distance from the sun as is Mars. The nominal doubling pattern repeats with Jupiter (780 million kilometers), Saturn (1,430 million kilometers), and Uranus (2,870 million kilometers).

The exception to the pattern of doubled spacing between the planets is earth. According to this pattern, there should be no planet in the exceptionally life-friendly region between Venus (too close) and Mars (too far). But here we are. As we say, the exception proves the rule. The rule here is that we reside on a very special planet at a very special location within a very special stellar system, formed at just the right position within the right kind of galaxy. The earth's distance from the sun, for the right amount of warmth, and its mass and gravity, for the ability to retain a proper atmosphere, put us in the only habitable zone within the solar system.[3]

Mars is already too far from the sun for tidal locking. It revolves around the sun in 687 of our days and rotates each 24.6 hours, a match of the earth's rotation rate. But at 228 million kilometers from the sun, insufficient solar radiation reaches it. Average surface temperature is minus 50 degrees C (minus 58 degrees F). Mars is also much smaller than the earth, with one-tenth the earth's mass and just over half the earth's diameter. Its gravity is too weak to hold a deep atmosphere. Surface atmospheric pressure is approximately 1 percent of earth's. Though the surface topography of Mars shows geological signs of past water-derived erosion, there are no clear signs of abundant liquid water on its surface today. Its weak gravitational field may have allowed the gradual loss of water vapor to space. If the locations of earth and Mars were reversed, there'd be a slim chance of life larger than one-celled microbes developing on either planet.

The next planet past Mars, or more accurately stated, the debris that might have formed into planet were it in another location, is the asteroid belt. That ring of stellar debris, rocks and boulders, might have coalesced into a planet, had it not been for the huge disruptive gravitational forces of giant Jupiter, the next planet in the solar system. The gravity of Jupiter working against the gravity of the sun literally tore apart any cohesive groupings of these asteroid rocks that might have formed into a planet.

The distance between the asteroid belt and Jupiter is similar to the distance between the asteroid belt and the sun. Had a planet the mass of Jupiter been able to form in the region now occupied by the asteroids, then, just as the gravitational force of Jupiter has obviated the possibility of a planet forming at that distance, the gravitational force of the hypothetical Jupiter-like planet located in the asteroid belt would have torn apart any inner planets between it and the sun, the earth included. There'd be no planet within the habitable zone of the solar system. Are we seeing the hidden hand of God at work here? Perhaps.

Though Jupiter's massive gravitational field halted the formation of a planet from the asteroids, its presence along with the other massive gaseous planet, Saturn, is essential for the complex life found on today's earth. Jupiter, at over three hundred times the mass of the earth, and Saturn, at just under a hundred earth masses, plus the other two large outer planets, Uranus and Neptune, contain 99 percent of the total mass of the solar system with the exclusion of the sun. This huge concentration of mass, and hence of gravity, in the outer regions has worked wonders in helping to rid the inner planetary regions of extraneous debris.

The pock-marked surfaces of the moon and nearly atmosphere-free Mars give evidence of a massive bombardment of asteroids during the earlier phases of the solar system. The earth was not immune to this devastating assault. Though highly eroded by our active weather systems, craters evidencing this blitz are still found on the earth's surface. Two hundred and fifty million years ago what appears to have been a massive impact on the earth led to the destruction of 90 percent of ocean species and 70 percent of all land species. It was decimation in the fullest literal meaning of that word, marking the end of the Permian age of forests of ferns and oceans teeming with trilobites. Trilobites were among the earliest of the insects, appearing in the fossil record at the Cambrian explosion of animal life, 530 million years ago. That era marked the initial appearance of animals larger than microscopic in size. And at their debut trilobites already had multifaceted insect eyes. Recall that Darwin said that the evolution of eyes gave him nightmares, so complex is that organ. Yet nature somehow (somehow?) already had eyes functioning at the inception of macroscopic animal life.

Sixty-five million years ago, a meteor estimated to have been 10 kilometers in diameter punched through the earth's atmosphere and exploded into the western Caribbean just off Mexico's Yucatan peninsula. Remnants of a crater 150 kilometers in diameter mark the site. Massive amounts of dust, jettisoned into the atmosphere, circled the globe. It's estimated that for half a year sunlight barely reached the earth's surface. Temperatures plummeted, photosynthesis all but ceased, and all life larger than a few kilograms disappeared from the fossil record. This marked the close of a 185-million-year reign of dinosaurs. Mammals had coexisted with the dinosaurs for most of that time, but they had remained diminutive. The small warm-blooded mammals survived the impact and flourished in the newly opened ecospace.

Meteorites burn in flashes of light as they stream through our upper atmosphere at 30 kilometers a second. They bear testimony to the reality that the formation of the planets and sun did not incorporate all the cosmic remnants of the previous supernovae from which we formed. The gravitational "vacuuming" by our giant outer planets removed vast amounts of rocky debris left over from those stellar explosions. Without this cleanup, the earth would have seen much more devastation. Quite possibly we would not be here to plumb the sources of our origins. Our primordial ancestors would have fallen prey to those annihilating blasts.

In July 1994, Jupiter's protective power was filmed in detail as the massive Shoemaker-Levy comet was pulled from its orbit and crashed into the planet. The Hubble space telescope captured the event, as the immense force of Jupiter's gravity first shattered the comet and then dragged the pieces into its gaseous body. The largest fragment produced an impact site that exceeded the size of the earth. It does not require much of an imagination to conjure up the image of what would have happened had that scenario taken place on earth. Our solar system has a very special design regardless of whether or not it had a Designer.

The right sized star, located between spiral arms, in a region of few stars, with an exceptionally high abundance of heavy elements needed to form the basis of rocky planets and life, having inner planets one of which has just the correct mass and gravity to hold an oxygen-rich atmosphere and is in the zone where the star's radiant energy can keep water in its liquid phase year-round, with giant outer planets sweeping up potentially life-threatening meteors and comets. The earth and sun certainly drew a royal flush and then some when the cards of life were being dealt. But there's more.

We've got tilt, axis tilt of 23.5 degrees off of vertical. It may not seem like much, but it spells wonders for moderating the climate at all latitudes. A vertical axis means constant year-round direct sunlight at the equator and only minimal glancing sun at both poles. The tilt of our earth ensures winter and summer seasons reach essentially the entire globe. Ocean and atmospheric currents further moderate the earth's climate. Under the influence of what is termed the Coriolis force, that is, the effect of a globe's rotation on the flow of a fluid over that globe (the ocean currents on the rotating earth), the west-to-east rotation of the earth induces clockwise ocean currents in the Northern Hemisphere and counterclockwise currents south of the equator. This flow, such as the Gulf Stream in the Atlantic and the Humboldt and Kuroshio (Japan) currents in the Pacific, moves masses of warm equatorial waters poleward and cold polar waters toward the equator. The net result is a distribution of warmth and a corresponding major increase in the regions amenable to habitation.

Until a draught comes along, we tend to take water as a given fact of daily life. The vast water reserves of the earth were brought to it by the ice-covered stellar debris that coalesced during its formation. Part of that water was lost to space in the early phases of our planet, but enough remains to cover totally a smooth earth to a depth of some two miles. Life might have developed on an earth totally covered by oceans. Certainly exotic and varied communities of plants and animals live around the thermal vents that mark the fissures of our ocean floors. There is, however, strong indication that advanced, intelligent life requires a solid and dry footing. Even whales and porpoises, the most intelligent marine life known, appear to have had a land-based ancestry that then migrated to the sea. Continents and the challenges they present may be an essential ingredient in the recipe for intelligent life. And to form continents we need a planet with a molten core and a solid crust.

During the meltdown of the young earth, the heavier elements, especially iron, migrated toward the earth's center. In time the slightly lighter minerals on the surface of the earth cooled and formed the rocky crust. The deeper layers remained molten, their heat being contained by the insulating effect of the overlying solid crust and added to by the energy release of several long-lived radioactive nuclides, most especially uranium, thorium, and potassium. Ever so slowly, the earth's molten interior overturns in huge convection cells made of viscous iron-rich magma. The hotter lower layers rise toward the crust, slide along the lower surface of the crust, and in doing so lose a small fraction of their heat. Then, being cooler and therefore now slightly more dense, they sink toward the hotter earth center.

The friction, as those molten thermal cells push along the overlying crust, has broken the earth's crust into continent-sized blocks. These blocks slide over the earth's surface at about a centimeter a year. The leading edges buckle as they force their way forward, rising in much the way snow does against the forward push of a plow. And in the manner of a snowplow, some of the encountered seabed slips under the continental block, sending the trapped elements deep below the surface, far from the biosphere. The rocks that rise form the mountain chains that line the western coasts of North and South America as these massive continents slowly push westward. The towering Himalayas result as India thrusts northward into Asia, and the Afro-Syrian rift, which isolated the eastern third of Africa from the western two-thirds of that continent eight million years ago and so heavily influenced the development of mammalian life, finds its source as two tectonic plates work against each other. That rift continues northward through Aqaba/Eilat, the Dead Sea and Jericho, eventually turning eastward into Syria.

The resulting mountain chains produce highly diverse climates, wet upwind of the mountains and dry downwind. As the air rises to pass over the mountains, it cools due to its expansion, and the air's moisture precipitates as rain or snow. Diverse climates engender diverse and hearty biospheres, as environmental challenges have to be met and conquered. The ratio of two-thirds ocean area to one-third continental, which the earth enjoys, has worked well for life. Weathering erodes the continental blocks carrying nutrient-rich sediment to the seas. The huge ocean currents distribute the solar input of heat over the globe. The exceptionally high specific heat of water (the amount of heat required to induce a given change in temperature) is greater than for most other common liquids. This trait combined with water's uniquely high latent heat of evaporation (the amount of heat needed to change water from a liquid to a gas) helps buffer against severe sudden changes in atmospheric and ocean temperatures.

Of course if a little is good, a lot is not necessarily better. Many asteroids contain one hundred times the concentrations of carbon and minerally bound water than the earth. Had the earth retained that composition, an ocean with a depth in excess of 10 miles would have originally covered the entire globe. The vast amount of carbon would have funded a carbon dioxide–rich atmosphere. The atmospheric carbon dioxide, CO_2 plus the added water vapor (H_2O) from the oceans, would have led to a runaway greenhouse effect, the result of long-wave energy being trapped by atmospheric gases with three or more atoms.

The planet Venus gives us a hint as to what that would mean. Venus's diameter is so similar to the earth's that it appears as a sister planet: 12,100 kilometers versus 12,800 kilometers for the earth. Being just under 30 percent closer to the sun than the earth (108 million kilometers versus 150 million kilometers), Venus receives more solar energy. But its 480-degree C surface temperature (lead is molten at that temperature) is not due to its proximity to the sun. The heat-retaining dense carbon dioxide atmosphere that shrouds the planet is the culprit. For example, compare the fact that the sunlit side of atmosphere-free Mercury is "only" 430 degrees C. That is 50 degrees C cooler than Venus.

Yet, at 58 million kilometers from the sun, Mercury is half the distance of Venus to the sun. The uncontrolled greenhouse atmosphere of Venus obviates its possibility to host life. How fortunate that the earth lost most of its carbon and water in its early stages of formation.

The characteristics of the sun's light are in themselves awe-inspiring. Approximately 70 percent of the emitted solar radiation falls within a wave-length range of 300 to 1,500 nanometers. This narrow band is fixed by the sun's surface temperature and so is not surprising. Radiation detected as light by a human eye falls between 400 (seen as blue) and 700 (seen as red) nanometers. What is surprising is that the gases of our atmosphere, oxygen, nitrogen, carbon dioxide, and water vapor, are transparent to this narrow band of incoming solar radiant energy while highly absorbent to both ultraviolet (shorter) and infrared (longer) radiations, those bands of energy on either side of the visible spectrum. Because of this match, the incoming beams of sunlight pass freely through the atmosphere, and so are available to power the essential processes of photosynthesis. It is photosynthesis that lies at the base of all food chains and also, as a side product, has liberated the oxygen that fills our life-supporting atmosphere. It might have been different. Never take the God-given munificence of our home in the universe for granted.

The uniqueness of this fit is reminiscent of the one-of-a-kind behavior of water expanding just before reaching its freezing temperature. Almost all other liquids contract upon changing from liquid to solid. That contraction makes them denser than the liquid from which they solidified (froze). As such the solids that form sink within the parent liquid. Because water expands as it freezes from water into ice, the ice is lighter than the water from which it froze. Being lighter, it floats. If it were otherwise, oceans would freeze from the bottom up, a reality that would in time bind most of the water permanently at frozen ocean bottoms. No abundant supply of liquid water means no advanced life.

Notwithstanding the late Carl Sagan's claim in his *Cosmos* series that our solar system is "ordinary, even mediocre," and Preston Cloud's statement in his widely quoted book *Oasis in Space* that our sun is "an unexceptional main-sequence star," our placement in the cosmos is anything but pedestrian. So special is our home in the universe that in what is considered by many to be the most authoritative geology textbook, *Earth,* written by Frank Press (of the National Academy of Sciences and formerly head of geology at M.I.T.) and Raymond Siever (of Harvard University), the opening section is titled "The Uniqueness of Planet Earth." That says it all, though Press and Siever expand upon this uniqueness in the following pages.

In order to begin to get some hint as to how many planets or even moons able to support such life there might be in our universe, let's review each of the variables needed for advanced intelligent life on a planet circling a star:

1. a spiral galaxy not in collision;
2. a low concentration of stars in the local region of the target star (most stars are within the high-density galactic center and spiral arms; many are binary stars);
3. location in the galactic region of high local density of metals to produce a metal-rich stellar system;
4. a star with a mass similar to the sun's mass, so that its energy output remains nominally constant for at least five billion years;
5. formation of a stellar system around the remnants of a second- or third-generation supernova (a second- or

third-generation supernova is required to account for the needed heavy elements that formed in the previous supernovae) within a few billion years following the most recent supernova (to ensure that sufficient radioactive uranium, thorium, and potassium remain to supply the radiant heat required to maintain the thermal cells of the iron core in their molten state, the motion of which produces, among other effects, the magnetic field of the earth, a force that shields the earth's surface from much lethal cosmic radiation);

6. a metal-rich planet (to have the elements needed to form the complex molecules of life);
7. a low (relative to asteroid composition) carbon content (to avoid a heat-trapping, carbon dioxide–rich atmosphere);
8. a low (relative to asteroid composition) water content (to allow moderate ocean cover, but also the presence of dry land masses);
9. a rotating molten iron-rich core in the planet (to induce a radiation-shielding magnetic field);
10. a planet with mass able to hold an oxygen-rich but not hydrogen-rich atmosphere (hydrogen is too chemically reactive);
11. plate tectonics (to produce continental drift and the resulting mountain chains yielding diverse climates and therefore a diverse biosphere);
12. a planet with nearly circular orbit (to ensure fairly constant solar input year-round);

13. a planet at approximately 150 million kilometers from a sunlike star (close enough to receive sufficient sunlight for warmth and oxygen-releasing photosynthesis, but not so close as to tidally lock it to its star);
14. a planetary system with huge outer planets (to help reduce the frequency of devastating meteor impacts on the inner planets; of the thousands of stars studied in the search for stars with planets circling them, to date only a few have been found to have planets and almost all of those to date have been massive and close to the parent star; the planets are observed by the periodic wobbling of stars or by light occlusion as the planet crosses our line of sight to the star);
15. a planetary system with no huge inner planets (their gravitational interaction with smaller inner planets would either obviate the formation of coherent inner planets as exemplified by the asteroid belt, or if planets did form, they would gradually be driven into the central star);
16. planetary rotation with a period on the order of days (to ensure all-around distribution of incoming solar radiation);
17. a moderate tilt to the planet's axis of rotation (to induce a more uniform north-south distribution of incoming solar radiation); and
18. a large moon (for tidal mixing).

Notice that none of these variables attempts to estimate how likely or unlikely it is that life could have formed from the rocks and water and few simple molecules that were characteristic of the early earth or of any young planet. There are no data upon which to base such a calculation. It would be pure speculation. Also we take as givens the existence of a rotating universe with stable laws of nature that provide a reproducible physics and chemistry and with three spatial dimensions and one time dimension. There's no attempt to ask how unlikely it is for chance to have constructed such a universe with just the correct physics. *Scientific American* already went that route and inferred that if our universe is the only universe, then indications are that we are a designer product. Our universe and our planet are both very special.

Let's make some guesstimates of the likelihood of finding in our universe each of the variables listed above. We can base our calculations on cosmological observations of the other galaxies, the distribution of stars and planets in our galaxy, the Milky Way, and the distribution and characteristics of planets in our solar system:

1. The frequency of isolated spiral galaxies is about one galaxy in ten, or 0.1.

2. The fraction of stars in the galaxy located in a region of low local density of stars, that is, between the spiral arms and far from the galactic center, 0.001.

3. The fraction of stars in the galaxy located in a region of high local density of metals, 0.01.

4. The fraction of stars with a mass similar to the sun's mass, 0.05.

5. The formation of a stellar system around the remnants of a second- or third-generation supernova within a few billion years of the most recent supernova, 0.1.

6. The fraction of planets that are metal-rich planets, 0.5.
7. The fraction of planets that have low (relative to asteroid composition) carbon content, 0.3.
8. The fraction of planets that have low (relative to asteroid composition) water content, 0.3.
9. The fraction of planets that have a molten iron-rich core, 0.2.
10. The fraction of planets with mass (gravity) able to hold an oxygen-rich but not hydrogen-rich atmosphere, 0.2.
11. The fraction of planets with continent-forming plate tectonics, 0.1.
12. The fraction of planets with a nearly circular orbit, 0.5.
13. The fraction of planets approximately 150 million kilometers from a sunlike star, 0.1.
14. The fraction of planetary systems with huge outer planets, 0.01.
15. The fraction of planetary systems with no huge inner planets, 0.1.
16. The fraction of planets with a period of planetary rotation on the order of days, 0.5.
17. The fraction of planets with a moderate tilt to the planet's axis of rotation, 0.2.
18. The fraction of planets with a large moon, 0.05.

Multiplying these admittedly gross, but nonetheless conservative, estimates, which are based on the conditions observed within our life-supporting galaxy and solar system and on our life-nurturing earth, the likelihood of finding anywhere in the entire universe a stellar system with a planet able to support complex intelligent life is one chance in 10^{18}. The estimated number of stars in the entire visible universe is in the order of 10^{22}. This indicates that in the entire universe there may be approximately 10^4, or 10,000, earthlike planets circling a sunlike star. These 10,000 potentially earthlike planets would be distributed among the 10^{11}, or 100,000,000,000, galaxies in the entire visible universe. That comes out to be one earthlike planet for each 10,000,000 galaxies. The probability that any one galaxy would have more than one life-bearing stellar system is slim indeed.

Of course, this estimate relates only to the possibility of finding conditions that would nurture life once it started. We learn nothing here concerning the properties required for the inception of life. Since we have no knowledge as to how life started on earth, and started remarkably rapidly, there are no data on which to base an estimation that might indicate the likelihood of life originating on any one of those rare life-friendly planets. Considering the uniqueness of our home in space, chances are that we are alone in our galaxy and possibly alone in the universe.

One fact is certain. We live on a very special planet that circles our star in a very special way, in a universe that has all the markings of being the product of design. In our study of nature, we've also discerned statistical evidence for God's intervention in the development of life. The Bible tells us that God is also active on a far more local scale. We read that God has made the land of Canaan to flow with milk and honey (Exod. 3:17) if we behave ourselves, and that God watches over this land throughout the entire year (Deut. 11:11). The land of Canaan is not the only land of interest to God. Other nations have been given other lands (Gen. 10:5–6). The generalized message in all these observations, from both the Bible and nature, is that God, the Creator, is intimately involved with and active in the creation It brought into being.

In the following chapters the goal will be to identify the characteristics of those Divine interactions. For me the discovery in this search was that, according to the Bible, God's actions in this world are vastly different from what I had assumed they would be based on the usual version of God as a Force that is omniscient, omnipotent, and totally in charge.

FOUR

Nature Rebels

God Grants Nature a Mind of Its Own

"And God saw all that He had made, and behold it was very good. And there was evening and there was morning the sixth day" (Gen. 1:31). This, the closing verse of the first chapter of Genesis, the celebrated, evocative creation chapter of the Bible, summarizes the results of the first six days of the universe: all was "very good." But was it all really "very good"? I wonder by what standards it was very good.

As we discussed in the previous two chapters, the universe had been meticulously made ready for life. But life, especially human life, seems not to have been ready for the responsibilities assigned to it. God had created Adam and Eve and "blessed them . . . and placed them in the Garden of Eden . . . saying, of every tree in the Garden you must eat, but of the Tree of Knowledge of Good and Evil you shall not eat" (1:27–28; 2:15–17). Eden was literally paradise on earth. Except for the one forbidden fruit, everything should have been "very good." Yet just two chapters after this heartwarming news of all being good, Adam and Eve rebelled against God, ate from the forbidden tree, and were expelled from Eden. A few verses farther on, Adam and Eve's firstborn murdered their younger son. Not just simple murder, but fratricide! That doesn't sound very good to me. A mere two more chapters pass, and God, with "a saddened heart," disgusted with the decay of society, throws in the towel and brings on the Flood to destroy all life, for God "regretted having made them" (6:7). Sounds like things weren't "very good" by God's standards either.

If God is supposed to be so great, couldn't God have controlled events a bit more strictly or at least realized that the entire project was a no-go right from the start? What was the Author of the Bible—and even more than the Author of the Bible, the Creator of the universe and life itself—thinking about when informing us that all was "very good"?

Henny Youngman, the late gifted comedian, said that after reading about the evils of alcohol, carefully reviewing all the documentation on the subject he could find, he decided to give up reading. That's called cognitive dissonance. I see the facts, but I've got my own ideas, so please don't confuse me with those facts. I prefer my personal take on reality.

I want a God that acts in a way I assume God should act, predictable according to my human logic. When the God of the Bible tells me all is very good, I expect all to be good. But then the Bible tells me not to count on it and lists disaster after disaster. So why was I told just the opposite? It's a hard truth to accept, and in our typically human desire to resolve cognitive dissonance, we argue all the way. But what we learn in the very opening chapters of the Bible is that the God of the Bible is not a predictable, static Divinity. That conventional but ill-conceived description totally misses the biblical reality.

We discover the startling truth of God's character in Exodus, the second book of the Bible. Exodus 3:14 is a verse often mistranslated and yet pivotal in understanding God's sometimes less than manifest immanence in the world It created. Moses, having been confronted by God at the burning bush, asks God's name (3:13). In reply, "God said to Moses, 'I will be that which I will be' [*ehe'ye* (I will be) *asher* (that which) *ehe'ye* (I will be)]. . . . This is My name forever" (3:14–15). This meaning of the Hebrew text is vastly different from the King James rendering of that verse, "I am that I am." The erroneous King James version (ca. 1611) is based on the fourth-century Latin Vulgate, which in turn was based on the six-hundred-years earlier Greek translation (the Septuagint). The irony of this ongoing error is that the exact Hebrew word in question, *ehe'ye,* appears just two verses earlier, in Exodus 3:12, and both the Latin and the Greek translations render this "I will be," not "I am." But "I am" is so much more predictable, more appealing to our preconceived notions of God than "I will be" that the translators actually changed the meaning of the biblical text!

Why belabor the point? Because what we discover here is that the God of the Bible is not a static Divinity, able to be pigeonholed into how *we* think God should act. *Ehe'ye* is not a present-tense verb or a noun, with all the implications of stasis, but a genderless verb actively projecting into the future. As we will learn in the following pages, the God of the Bible is a dynamic Force with options, contingency plans, a manifestation that changes to fit the changing needs of the dynamic world It created. *Ehe'ye* is the perfect description of the God of the Bible. We expect an "I am," but instead the God of the Bible self-identifies as "I will be."

If we are to treat the Bible as a valid source of information about God's role in our lives, then looking closely at the text is a prerequisite. As King Solomon urged in Proverbs (25:11), we must seek the "apples of gold in the dish of silver"—the deeper truths sequestered within the literal text. To help us do that, throughout this book I will be using the major ancient Hebrew commentaries.

Among the oldest of these works is the Talmud (redacted ca. 400). It is a compendium of biblical commentary and exegesis from the four-hundred-year period prior to its redaction. Most of the text is in the form of debates between scholars seeking the correct interpretation of the Bible's wording, which, as we've seen, can be very subtle. Rashi (1040–1105), a scholar considered to be the interpreter of biblical Hebrew par excellence, lived in southern France. He was also a vintner. Yet for all the demands of the wine industry, Rashi was able to compose precise commentaries on much of the Talmud, a text having over five thousand pages, plus the primary Hebrew commentary on the Torah. The philosopher and theologian Maimonides (1135–1204), while living in what is now Egypt, was chief physician to the ruler of that country (a post that had its very real dangers). He compiled a codification of the laws found in the Torah with a commentary that exceeded nineteen volumes. An example of his philosophical work is *The Guide of the Perplexed* (1190), which deals with many aspects of the Bible that seemed to conflict with how the world was perceived. Nahmanides (ca. 1195–1270) was a kabalist who offered insights about the Bible that he learned from his teachers. The Hebrew root of the word "Kabala" means "to receive." The received wisdom of Kabala is considered to be the spiritual physics of the world, essentially how an infinite eternal nonphysical Creator interacts with the finite physical world It created. Kabala itself is not mysticism. However, the sensation one may get when internalizing the information can lead to what might be called a mystical experience.

There are many commentaries and commentators on the Bible. These are at the top of the list. Being ancient, the perspectives that they bring are not biased by the discoveries of modern science. The antiquity of these sources ensures that there has been no attempt to bend the biblical text to match science. In this they offer us a unique and invaluable vantage point. And for that reason, only ancient commentaries are used in this book as authoritative sources of biblical insight.

The first six days as described in Genesis 1 take us from the creation of the universe to the creation of humanity. The entire account is described in a mere thirty-one verses. At M.I.T.'s Hayden library, we probably have twenty thousand books on the events covered in those six days—not from a theological perspective, but from a scientific one that deals with the cosmology, physics, and biology of a universe created with energy capable of producing life, brain, and sentient mind. Up the Charles River from M.I.T., at Harvard's Weidner library there are probably fifty thousand books on these topics. With only thirty-one verses in the first chapter of Genesis, we shouldn't expect each Divine detail in the cosmic development described there to leap off the biblical pages. If we are to reap the golden apples within the silver dish of the Bible, we'll have to reach out. It is worth the effort.

The flow of events during the first six days of creation is driven by the recurring Divine command "And God said . . ." This directive, which appears nine times in the first chapter and a tenth in the second, presages the unfolding drama of a universe in the making. With God so intimately in control we should expect smooth sailing in that most amazing of cosmic voyages. Surprisingly, according to the biblical account of those events, that seems not to have been the case. The very fact that in those thirty-one verses the Bible felt it necessary to enlighten us seven times over that God saw "it was good" and even "very good" might imply that perhaps at times it was not so good. Of course that would be absurd according to the image we project of an infinite, always-in-control God, the father figure of the Bible, creator of the heavens and the earth. But let's see what the Bible has to say.

On the third day of the creation, we are told that God commanded the earth to bring forth the first forms of plant life, vegetation. It's interesting that the word "creation" does not appear on day three. Creation would signify that something entirely new, an entity unable to be made from the materials already present, was needed. No mention of creation means that nothing totally new was needed to bring life into the universe. The big-bang creation produced the physical basis for all the materials required for life. This inherent potential for life to flourish on earth led Nobel laureate, organic chemist, and authority on origin of life studies Professor Christian de Duve to write in *Tour of a Living Cell:* "If you equate the probability of the birth of a bacteria cell to chance assembly of its atoms, eternity will not suffice to produce one. . . . Faced with the enormous sum of lucky draws behind the success of the evolutionary game, one may legitimately wonder to what extent this success is actually written into the fabric of the universe."[1]

What was needed to spin out the fabric's potential for life was the Divine command, "And God said . . .": "And God said let the earth sprout vegetation, herbs yielding seed, fruit trees yielding fruit each after its own kind with its seed in it . . ." (Gen. 1:11). That verse is the statement of the Divine command for the earth to produce the first forms of life. The description of nature's execution of the command follows in the next verse. "And the earth brought forth vegetation, herbs yielding seeds of its kind, and trees yielding fruit with its seed in it after its kind . . ." (1:12). The execution of the command seems essentially the same as the command itself. Seems the same, that is, until we read the words more closely with the help of Rashi's decisive ancient commentary. God's command asked for "fruit trees yielding fruit," but the earth produced "trees yielding fruit."

A minuscule, seemingly insignificant divergence, but an astonishing implication is revealed to us by Rashi. Fruit trees yielding fruit has a superfluous adjective, the word "fruit" modifying "trees." The text might have simply stated "trees yielding fruit" or, equally descriptive, "fruit trees." That would have been sufficient to indicate the directive for the earth to produce trees bearing fruit. The ancient commentaries, upon which I am basing the intention of the biblical text, accepted that seemingly superfluous words, especially when presented or omitted in successive verses such as here in verses 11–12, come to bring one of the "golden apples in the silver dish" of the Bible. Rashi wrote that the earth was commanded to produce fruit trees yielding fruit in order that the taste of the tree or its bark as well as the fruit that hung from its branches would have the taste of a fruit. But the earth did not comply. *The earth rebelled.* Instead, the earth brought forth trees bearing fruit, not fruit trees bearing fruit, the wood or bark of which would also be as a fruit.

Perhaps God's demand exceeded nature's potential. Can the wood of a tree ever be a fruit?

In the family orchard of my parents, of blessed memory, we had mostly apple trees, plus several quince, peach, and pear trees, and two cherry trees (which year after year the birds harvested before we could). Though I occasionally chewed on a branch as I was working at the yearly pruning, regardless which of those trees it was, the wood always tasted like wood and never like the fruit of that tree. Except for the cinnamon tree, the bark of which is the "fruit," hence making it a "fruit tree," we might have thought that such a thing as "tasty wood" would be an impossibility. Not so. Nature simply failed to comply with God's command.

And even the cinnamon tree did not succeed in completing the Divine command. Although it is indeed a "fruit tree" because its bark is a spice, it does not also yield a separate fruit to become "a fruit tree yielding fruit."

The Torah didn't have to tell us of this rebellion. If nature's mutiny had been kept a secret, all we would know is trees of wood. But the Bible comes to teach reality, not some fantasy we might have desired of a story-book God orchestrating a make-believe world. Rashi makes a point of informing us about this rebellion to break that image of an always controlling Master, a superpowerful father image. Such a God makes a lovely, heart-warming, even reassuring fable, but unless the Bible got it all wrong, the ever hands-on God of our childlike imaginations is not the Deity of the Bible active in our world.

The Bible in these verses tells us an almost incomprehensible fact. Nature, purportedly bound by unbending "laws of nature," which were themselves created by God, somehow was able to do the unimaginable. Nature was able to go against God's explicit command. Nature rebelled. If someone of less stature than Rashi had made this biblical "accusation," it would have been dismissed as trivial nit-picking. But Rashi is the foremost of all Hebrew commentators on the Bible.

And more intriguing, this three-thousand-year-old biblical claim of nature rebelling against the command of God, of nature having a "mind of its own," is eerily similar to modern-day observations of quantum physics. As noted previously, Sir James Jeans, knighted mathematician and physicist who helped develop our understanding of the evolution of stars more than seventy years ago, wrote in his book *The Mysterious Universe:* "There is a wide measure of agreement *which, on the physical side of science approaches almost unanimity,* that the stream of knowledge is heading towards a non-mechanical reality; the universe begins to look more like a great thought than a great machine. Mind no longer appears as an accidental intruder into the realm of matter. We are beginning to suspect that we ought rather to hail mind as the creator and governor of the realm of matter" (emphasis added).[2]

Does this imply that something akin to mind exists throughout all of nature, some phenomenon along the lines of Nobel laureate Wald's previously quoted epiphany? And that this "mind," which of course finds its origins in the Divine creation of the world, could have allowed the earth to rebel? Sounds preposterous. To the uninitiated. But not to those who struggle to conceptualize the quantum reality of existence. To what level of existence does mind, and all the implications of what mind brings, including a level of self-awareness, extend? Humans of course have it. What about dogs and cats? Pass a few hours in any public park, and from the way dogs and cats react you realize that dogs know that they are dogs and not cats. But these are all mammals. What about other forms of life? Do they too have mind?

While driving in Holland with my family, we noticed a huge nest mounted on what appeared to be a utility pole. I asked a local resident and was told it was a stork's nest, occupied each summer by what appears to be the same bird as it completes its annual multithousand mile migration from northern Europe in the fall to sunny Africa and its return as the seasons reverse. I'd heard of that navigational feat. But until I met naturalist Rabbi Shmuel Silinsky I'd never internalized what a navigational achievement this represented. Just ponder, he said, how many rooftops and utility poles there are in Holland. I couldn't find our way without a map and road signs, but the stork's bird brain does it just fine, as do monarch butterflies in their flight from Mexico to Alaska and back, even though their migration takes several generations in each direction.

A review article in *National Geographic* surveyed the level of cognition within a range of animals. Scientists from such esteemed universities as Stanford, Duke, Harvard, and Brandeis describe the extraordinary level of awareness, including self-awareness and episodic memory (the cognitive ability to mentally travel back in time), in animals as different as elephants and birds.[3] Parrots, Irene Pepperberg of Harvard and Brandeis points out, are especially interesting because parrots can learn to speak. Her colleague during thirty-one years of research, Alex, an African gray parrot, did exactly that, identifying shape and color and numbers, even asking for breakfast and a view out of a window in another room—not in symbols, but in English words.

Animals have it, but what about microbes? Do they communicate in a meaningful way, a way that can actually transfer information? If we are claiming that mind, rebellious or not, exists in the seemingly inert earth itself, then we should be able to identify its presence in the most basic forms of life.

In the treatment of several recalcitrant microbial infections, vancomycin is often the antibiotic of last resort. Its powerful sword beat back a brutal illness, extending the life of my late father-in-law, of blessed memory. Bacteria that can outsmart many of our medical weapons often fall prey to this brilliant product of biotechnology. Vancomycin does its work by attacking the cell's first line of defense, the cell wall.

The bacterial cell wall differs from that of multicellular life-forms in that it has an external scaffolding that helps the single-celled critters maintain their shape. The structure of cells in multicellular life, such as humans, is established by a system of internal (rather than external) microtubules. In that difference lies the basis of vancomycin's effectiveness. It disrupts the construction of the bacteria's cell wall while not affecting the cells of the multicellular life-forms, such as humans, who might happen to be the unfortunate hosts of the infecting microbes. What the bacteria had to do to thwart the vancomycin antibiotic attack was to change ever so slightly the molecular structure of their cell wall, so that the vancomycin could no longer recognize and attack them. And change they did.

The chosen alteration is brilliantly subtle and eminently effective. The bacteria replace one of the usual cell-wall amino acids, alanine, with lactic acid, a molecule sufficiently similar to the alanine so that it fits easily into the structure, but sufficiently dissimilar so that the vancomycin can no longer do its job. Lactic acid substitution is a logical choice. It's already a common metabolic product, being produced as an end product when the digestion of glucose occurs in the absence of oxygen. So there's nothing new about the presence of lactic acid in a cell. The sore muscles we get when overexercising result in part from the accumulation of lactic acid in our muscle tissues. As with alanine, the lactic acid molecule has three carbon atoms, one methyl group, and one double-bonded oxygen atom. The significant change is that the amine group of alanine is gone.

To accomplish the alanine–lactic acid switch, the bacteria assembled a genetic unit referred to as a plasmid, a snip of DNA in this case housing nine separate genes. (If you are seeking the hint of a cunning stratagem within bacteria, here it comes.) One of the nine genes in this package codes for a protein, an enzyme that cuts the bonds holding the alanine in place. A separate gene encourages the bacteria to manufacture the needed lactic acid. A third gene codes for the protein/enzyme that bonds the lactic acid into the lacuna left by the now absent alanine. Of course this rather complex act costs metabolic energy, which means it costs the bacteria food. Not being profligate, bacteria have no desire to perform this operation any old time. Not at all. Two other genes in this wonder-packed plasmid allow the aforementioned chain reaction to occur *only* (get that, *only*) when vancomycin is present. The exact function of the remaining genes is not completely certain, but we can rest assured that junk they are not. The research related to this microbial method of communication has been reported in the leading medical journals.[4]

Having devised this most valuable package for survival, the bacteria can share the product with other microbes. There are several methods for this transfer. In the most "mind"-like method, the "parent" microbe extends a microbial tube that connects with the recipient microbe. The parent then replicates a snippet of its genetic material and passes it via the tube to the receiving bacterium.

And all that wisdomlike ability is stuffed into a bacterium approximately a millionth of a meter long and a third as wide. Observing that this level of "mind" extends to the simplest form of life makes the biblical inference that the earth has the trait of mind less of a leap. But then in the very first words of the Bible, we read, "With a first cause that was wisdom God created the heavens and the earth." All aspects of creation are imbued, actually permeated, with the potential for mind. Over three millennia ago, the Bible taught this truth in black fire written on white fire, provided that we read both the black of the written words and the white of the wonders of nature as one integrated whole.

It's hard to accept that a form of life as "simple" or primitive as a bacterium might actually exhibit traits of brainlike shrewdness. Not so for neurosurgeon Frank Vertosick, Jr. In his excellent book *The Genius Within,* he places reality before us:

> If I speak with admiration for these creatures [bacteria] it's because I've won and lost many battles against them. . . . Unlike non-clinical biologists like [the late] Stephen Jay Gould or Richard Dawkins, physicians like myself enter the

> competitive arena and do battle with supposedly unintelligent beasts like bacteria and cancer cells. Darwin observed the creatures of the world with a keen eye, but he never fought them one on one. For those of us who stare into the shining eyes of the world's predators, we know how cunning they are at what they do. . . . The genes and enzymes and bacteria in this saga are but cogs in the greater communal machine, the microbial mind.[5]

Freeman Dyson is a physicist at the Institute for Advanced Study in Princeton, New Jersey. According to Dyson (and others): "Atoms are weird stuff, behaving like active agents rather than inert substances. They make unpredictable choices between alternative possibilities according to the laws of quantum mechanics. It appears that mind, as manifested by the capacity to make choices, is to some extent inherent in every atom. The universe is also weird, with its laws of nature that make it hospitable to the growth of mind."[6]

These scientists did not live or practice their professions in a vacuum. When they published statements such as these, not only did these thoughts have to pass the editorial review, but also these scientists had to face their colleagues and defend their positions. In short, to proffer such a diversion from the conventional material myth, Vertosick and Dyson had to be secure in their evaluations of reality.

If mind, or wisdom as biblically noted, is indeed the inherent essence of all existence, then the "mind" in nature, which of course is the product of God, could conceivably rebel, deviate from the word of God. Hitherto I thought only humans had that ability. We call it our free will. Yet if the Bible's description of reality is valid, then apparently I learned wrong.

When Adam and Eve are reprimanded for having eaten of the forbidden fruit of the Tree of Knowledge in the Garden of

Eden, God also punishes the ground. "And to Adam He said, 'Because you listened to the voice of your wife and ate from the tree which I commanded you saying you shall not eat from it, cursed is the ground because of you'" (Gen. 3:17). There is a non sequitur here. Adam and Eve ate the forbidden fruit and God cursed the ground! Why punish the ground at this point? It was Adam and Eve who ate the fruit. The ground was totally passive here. I can imagine the ground pleading with God at this point: "What did I do? You, God, planted the forbidden tree in the Garden. You made it fantastically 'appealing to the eye' [Gen. 3:6]. They eat of the fruit and you punish me? Why punish me?" There is, however, a common theme between the eating of the forbidden fruit and the ground not fulfilling its calling. And that theme is rebellion. First, on day three the "inanimate" ground rebelled by not "minding" God's command, and then later in Eden the first humans rebelled by deliberately disobeying God's command. Rebellion was in the air, and God chose to nip it in the bud. But He didn't, did He? As we learn a few chapters farther on in the biblical text, free will and rebellion remained facts of existence.

We are learning that the "Lord" of the Bible does not at all fit the usual concept of an all-controlling "Lord." But the Bible has already made this clear by the two primary names it uses for God. *Elokiim* relates to God as made manifest in nature. As such, *Elokiim* is the only name used for God in Genesis 1, the creation chapter. There we read of the physical development of the world. We will discover that nature is not as firmly bound by unyielding "laws of nature" as we might expect. In Genesis 2, the events more fully describe interactions between God and Adam and between Adam and Eve. From this point on, in addition to *Elokiim*, we find the fundamental four-letter name of God, a word that has an approximate transliteration of the Hebrew as *Ja/ko/vah*. Of course the continuing paradox remains: both *Elokiim* and *Ja/ko/vah* are the one God that chooses to manifest Itself in very different ways. We were told: "I will be that which I will be."

Recalling that *Elokiim,* God's control as made manifest in nature, is the only name for God used in Genesis 1, the creation chapter, could nature's ability to rebel be the source of the rare negative genetic mutations that mar what might otherwise be a properly formed baby at birth or allow for cancer to run rampant as it does today?

The screening and repair mechanisms in cell structure and function approach perfection. With only a small Divine increase in the molecular skill of cell repair, no mutations would succeed. There would be no malformed children and also perhaps no cancer. Yet mutations, those that are not detrimental, are what allow different forms of life to develop. They play a crucial role in forming the nuanced variety we observe within any community of living organisms. Variety is more than merely the spice of life. Variety within a species allows that species to adapt to changes in the environment, an aspect of life so essential for its robust and vigorous flow. Mutations act as a two-edged sword.

God created our universe with its inherent ability to diverge from God's Divine plan. In the opening chapter of Genesis we are explicitly instructed by God to have dominion over the world (1:28). By giving us such authority, God has placed with us the responsibility to repair the errors brought about by the vicissitudes of nature. An example of success in fulfilling this communal responsibility is found in medicine. The mortality rate of childhood oncology less than a century ago approached 100 percent. Today and for much of the past decade, the survival rate exceeds 90 percent in leading hospitals. That success was the result of the combined efforts of many disciplines. God has designed a world in which we are in truth our brothers' and sisters' keepers. A less auspicious incident in the book of Joshua makes this absolutely clear.

After the Israelites had walked in the desert for forty years, Joshua was charged with the task of leading this people into the Promised Land. Jericho, a fortified city located just west of the Jordan River, confronted their entry. Because of the rampant abominations practiced by its inhabitants, God demanded that the city be conquered. Taking booty in any form was forbidden (Josh. 6). Unfortunately the temptation for the gold and silver overwhelmed the prudence of one warrior: "Akhan, the son of Karmi, the son of Zavdi, the son of Zerah, of the tribe of Judah took for himself from the prohibited treasure" (7:1). At the following battle with the city of Ay, though it was a small town, the Israelites suffered a severe loss when the army was totally routed: "And the men of Ay smote about thirty-six of them and chased them from before the gate as far as Shevarim and smote them at the descent" (7:5).

Joshua realized that such total defeat could not be explained merely in military terms. It had to be Divine punishment. To discover the cause of the calamity, Joshua cast lots, choosing, by the roll of a series of dice, first from among the twelve tribes, then family by family within the "chosen" tribe, and finally person by person. At each roll, the choice came closer to and finally rested on Akhan: "And Joshua said to Akhan, 'My son, . . . tell what you have done.' And Akhan answered Joshua and said, 'Indeed I have sinned against God. When I saw among the spoil a fine cloak and two hundred shekels of silver and a block of gold of fifty shekels weight, I craved them and took them. They are hidden in the earth within my tent'" (7:19–21).

One person erred, and many suffered. The Bible describes the beginning of all humanity as stemming from a single couple. Accordingly, in the world God designed, communal responsibility is not bounded by tribe or geography.

Could God as described in the Bible run every detail of existence? Could God, for example, have forced the ground to bring forth fruit trees bearing fruit? The conventional answer is obviously yes. The Creator of this grand universe must certainly be able to control every aspect of its functioning, down to each blade of grass. Yet, notwithstanding this common perception of God being infinitely powerful and ever in control, every indication in the Bible is that God has chosen not to control all events. Biblically, this lacuna arises by Divine fiat and not by Divine necessity. Whether by fiat or necessity, the fact of this Divine decrease in control and its effects on society remain.

When going to war, just prior to entering battle, the combatants are asked: "What man is there that has built a house and has not yet dedicated it? He should go and return to his house lest he die in battle and another man dedicate it. . . . And is there a man who is engaged to a woman and has not yet wed her? He should go and return to his home lest he die in battle and another man wed her" (Deut. 20:5–7). Lest he die in battle? Couldn't God protect these particular soldiers? We'll deal with accidents in a later chapter. Here, however, we see an aspect of reality far more fundamental: there is no guarantee of survival to the individual, even though only a few verses earlier, the soldiers were told, "Fear not . . . for the Eternal your God goes with you to fight against your enemies to save you" (20:3–4). God may fight for national survival. That notwithstanding, individuals cannot rely on a miracle to save them. God has the eternal option of stepping back and allowing nature and people to take their course. As God told Moses at the burning bush, "I will be that which I will be" (*Ehe'ye asher ehe'ye*).

The Divinely imbued autonomy at all levels of nature from earth to Adam,[7] narrow though it may be, provides the potential for paths to be followed that may be less than beneficial. Quite simply, events occur that God would rather not occur. With this comes the latent possibility for undeserved and unexpected tragedy and even evil to enter our lives. Can it be that God foresees the trouble in the offing and nonetheless allows it to occur? In general God lets us travel the route we seek.

Prior to entering Canaan, the recently liberated Israelites of the Exodus asked Moses to send scouts into Canaan to reconnoiter the Promised Land (Deut. 1:22). God realized this was not a good idea. What if the report the scouts brought back was disappointing? But the people really wanted it. And so when okaying the plan, "God spoke to Moses saying, 'Send for yourself men to scout out the land of Canaan that I give to the children of Israel' " (Num. 13:1–2). God was saying, "Send for yourself and not for Me. You are courting trouble." And trouble came. The scouts returned forty days later and reported that the land was great, but the inhabitants were too powerful to confront. The people rebelled. Their rejection of the land that God had promised resulted in their being subjected to their forty-year trek in the desert; one year for each day the scouts had been in the land. Did God foresee the rebellion? We cannot know. But the wording of the Bible makes it clear that God did not like the plan and hence, "Send for yourself." God allows us to traverse the course we choose even though it may not be the most opportune option available. When Moses asked God to appoint people to help in the task of guiding the Israelites, God approved of the plan and so it is written, "And God said to Moses, 'Choose for Me seventy men . . .' " (Num. 11:16). "Choose for Me as well as for you."

This granted latitude in Divine control is less of a surprise, once we become familiar with the biblical description of what creation actually engenders. Biblically, we see creation as something, the universe, arising from the totally nonphysical we refer to as the Eternal Creator. But that is our view from within the physical side of creation looking toward the metaphysical. From the opposite perspective, from that of the Divine metaphysical viewing the physical, the perception of creation is quite different.

Speaking through the prophet Isaiah, God defines the Divine act of creation: "I am the Eternal and there is nothing else beside me. I form light and create darkness, I make peace and create evil. I the Eternal do all these" (45:6–7). "I *form* light and *create* darkness." One of the metaphors for God in the Bible is light: "By Your light we see light" (Ps. 36:9); "You wrap Yourself in light" (Ps. 104:2). God, the source of all spiritual light, "creates" spiritual darkness by withdrawing some of the Divine light: "I *make* peace and *create* evil." One of God's names is Peace, *shalom, shlaimoot,* "wholeness," "harmony." To have evil, discord, in a world constructed of peace, some of that peace must be withdrawn. From God's vantage point, the act of creation, in Hebrew, *ba're'ah,* entails a lessening of God's manifest presence and control. Creation according to the Bible is God's spiritual contraction. In Hebrew the term to describe this Divine contraction is *tzimtzum,* which literally means "to contract" or "to withdraw," in this case a partial withdrawal of God's evident spiritual presence. In essence, God hides God's face. What once might have been a simple unified whole becomes multifaceted, moving in a multitude of paths, not all of which are necessarily spiritually compatible. *Tzimtzum* provides spiritual space for all aspects of existence as we know it.

The first creation, and hence the first *tzimtzum*, God's creating the heavens and the earth (Gen. 1:1), brought into being the physical world with its time-space-matter, a single fabric interwoven by the laws of nature. The laws of nature are indeed "laws." However, sequestered within is a quantum slack, a leeway in those laws that control nature. At the subatomic level, identical causes do not yield identical effects. That also is the message of the *tzimtzum* of creation. Einstein is quoted as having said in response to this quantum uncertainty that he could not believe that God played dice with the universe. Einstein was correct. God does not play dice with the universe, but God allows the universe to play dice.

According to all ancient Hebrew commentaries, the creation of the universe described in the opening verse of Genesis was the only *physical* creation. One physical creation is also the message of science. We call it the big bang. Everything that ever was or will be is made from the light of that first creation. Thus wrote the kabalist Nahmanides eight hundred years ago, and so teach the discoveries of cosmology today.[8] That creation shattered an undifferentiated whole and produced the potential for the variety we see in our world.

The second creation, and hence the second *tzimtzum*, occurred on day five of the six Genesis days (1:21) and relates to the creation of the animals. This was not the creation of their bodies—those were made from the already existing material—but the creation of the wholly ethereal *nefesh*, the soul of animal life. The *nefesh* gives animals a level of choice and motion not found in plants. Animals choose among foods, driven in part by instinct. They can learn, navigate through a maze, and make tools. But the *nefesh* is totally self-centered, driven toward maximizing pleasure, survival, and progeny. The world, according to the *nefesh*'s view, is there to be exploited for the self's own needs.

The creation of Adam, recorded on the sixth day, involves the *neshama*, the soul of human life. The *neshama* attempts to change fundamentally the drives of the human animal. The *neshama* realizes that a spiritual unity pervades and unites all existence. This unity will eventually be spelled out as the central concept of biblical monotheism, "The Eternal is One" (Deut. 6:4; Mark 12:29). The *neshama* apprehends this ultimate message. Each person has a window of choice from within which he or she decides. Just as no two people are identical, so no two windows of choice are identical. But every person's *neshama* evaluates every choice made to determine if an act will move the individual closer to or farther from that Unity, the Oneness of existence. Reaching that Unity is the ultimate pleasure of life.

Each act of creation during the six days of creation was a further *tzimtzum* by God, a further allowing of ever more freedom in the manner by which God's commands were executed. The earth, acting within this relaxation of control, could in a sense "choose" to produce trees bearing fruit rather than fruit trees that also bore fruit (Gen. 1:11–12). And at the other end of the scale of freedom, Cain could choose to murder his brother, Abel (4:8).

Whether, according to the Bible or science, there is an actual *consciousness* of "choice" at the level of complexity of the earth or a tree is moot. We don't speak the language of soil or plants. The autonomy inherent at the physical level of the quantum, while not proving the existence of free will, opens the possibility for the concept of choice even in the assumedly inanimate world of atoms and molecules. After all, it is the same protons, neutrons, and electrons in differing combinations that make up all material existence from earth to Adam. At some point along this gradation of complexity, manifest consciousness has emerged. The question is not "if" self-awareness can arise from a particular mix of these seemingly inanimate subatomic particles. We are living proof that it can and did. The problem is to identify at what level of complexity sentience and choice and mind come quantifiably online. By the time we reach the third of the creations, that of the *neshama* of humans, our free will is at such an advanced level that the Divine leeway of *tzimtzum* has actually granted us license to choose between life and death, that of others and even of our own.

"I call heaven and earth to witness with you today, life and death I have placed before you, the blessing and the curse. Choose life in order that you may live, you and your children" (Deut. 30:19). This is a peculiar admonition by God. Wouldn't we assume that people normally choose life? Apparently God did not think so. In the Garden of Eden, 2,448 years prior to this revelation at Sinai, Adam and Eve were confronted with the identical options. There, two sources of food were specifically offered: the Tree of Life and the Tree of the Knowledge of Good and Evil. Eating from the latter would bring upon them their spiritual death. In contrast, the Tree of Life, from which they were told that they must eat (the verb in the Hebrew is doubled, implying a direct demand), represented the source of eternal Divine life.[9] This first couple on earth to have the soul of humankind, the *neshama,* actually chose knowledge of good and evil over Divine life. In the two and a half millennia between Eden and the revelation at Sinai, human nature had not changed. God now had learned the disposition of His creations and therefore urged that we make the choice that on the surface would seem obvious, but apparently was not so: we should go for a dynamic meaningful life over the stagnation of death. In a world where subjective feelings of pleasant and not pleasant blur interpretations of right and wrong, the right path is not always obvious.

In the creative act of God's *tzimtzum,* this withdrawal of absolute Divine control, we discover the source of chance and choice within our world. And to our astonishment, this granted autonomy extends throughout all levels of existence. Considering God's reaction to the rebellion of the "inanimate" earth at the incident of the fruit trees, and much later to the murder of Abel by brother Cain, anthropomorphically speaking, one is led to wonder whether the scope of Divinely granted freedom was more than God had originally bargained for. Some online revisions in the Divine management of the world were in order. With the failure of God's contingency "plan A," in which the first humans were to be nurtured in Eden, "plan B" was inaugurated. Adam and Eve were expelled from the Garden. Perhaps being exposed to the outside world would be more instructive. Unfortunately that approach also didn't fare any better. Within ten generations, society had so fully deteriorated that it needed the washout of the Flood. Once again God changed the venue for life. Is this a Divine learning process or is it more accurately described as God's very essence, the core meaning of "I will be that which I will be"?

FIVE

A Repentant God?

How to Understand a God That Has Regrets

In addition to Adam's and Eve's flawed intellect as evidenced by their rebellion against God's command not to eat of the forbidden fruit, there appears also to have been a flaw in the physical design of humans. Prior to the Flood, life spans of the people listed in the Bible were measured in centuries; reaching one's nine hundredth birthday was the norm.[1] But nine-hundred-year life spans were not an effective way to run a world in which free will had been granted through the *tzimtzum*, the constricting, of Divine control. With such a long tenure of life, one might easily lose sight of purpose and settle for less than admirable goals. And that is exactly what happened.

The earth provided a most hospitable home for human life. But all was not rosy. As mentioned, fratricide marred the very first generation of children born to Adam and Eve. Things went from bad to worse, and by the tenth generation of humans, the time of Noah, evil was rampant. It was so bad that God determined to restart the experiment of life. The Flood of the time of Noah was about to engulf the civilized world of Mesopotamia. Now note that we're only at the sixth chapter of the Bible. We still have 181 chapters to go before finishing the Torah, and already the Creator found it necessary to reset Its own creation.

"And God saw that the evil of humankind was great on the earth. . . . And God repented that He had made humankind [literally, 'the Adam'] on the earth. . . . And God said, 'I will erase humankind [the Adam] whom I have created from the face of the ground from human to beast . . . for I regret that I made them' " (Gen. 6:5–7). The Hebrew word translated here as "regret," *nehamti,* equally carries the meanings of "repented" or "reconsidered," all of which have within them the essence of a "change of mind."

The implication of these words is nothing less than mind-boggling. God said, "I regret." The Creator of the entire universe, maker of the heavens and the earth, has regrets. As Rashi points out, every instance of this word, "regret" (*nehamti*), has the meaning of reconsidering what to do. The verse quoted is not an interpretation of the biblical text, some commentary expanding upon the meaning. The Bible presents this reconsideration as a direct quote of the Creator, the Eternal God. If indeed the Creator is omniscient, knowing what is in store even before it happens, what place is there for regret or reconsidering? Designs crafted by humans may result in faulty craftsmanship. Airplanes may crash. Nuclear reactors may leak and even at times explode. Ships may sink, and clothes deteriorate. No one expects human designs to be absolutely perfect. But I had always assumed that Divine design implied perfect design. This seems not to be the case.

Our naive view of the predictable, omniscient God once again requires disturbing modification. The Bible's description of how God manages the world It created is completely consistent with what we observe in the world. And both fit the self-description of a dynamic Presence, varying Its manifestation to fit the varying reality of a world intrinsically imbued with free will. We know God through God's actions. "I will be that which I will be" is the total truth of God.

Sociologist Avner Mizrahi suggests that, considering the multiple failures of humankind, society might have fared better had God waited to select a candidate more fitting than Adam to receive the *neshama*. Even if there is leeway in God's control, the *tzimtzum* of creation, and even if Adam's progeny were less than optimal, why didn't God act to stem the evil before it became rampant? But that appears not to be the method of Divine management in our universe.

"Has the Eternal God's hand ever waned short?" (Num. 11:23), God chided Moses. The recently freed people craved for a food more substantial than their daily supply of manna from heaven. They wanted meat, and God told Moses that He would bring to the multitude of people enough food for a month and more. Moses was incredulous: "If flocks and herds were slain for them, if all the fish in the sea were gathered, will it be adequate?" (11:22). At which point God reminded Moses that God was ultimately in charge. His hand had not in the past and would not in the future "wane short." Yet from our human perspective, the need for a flood to revamp the world would imply that indeed something had "waned" very "short" in the design and implementation of our universe—so short that God "regretted" having made humankind. Something went awry, so painfully awry, that civilization had to be destroyed by the very Force that created it.

"And God regretted [or reconsidered] that He had made humankind and it grieved Him in His heart. . . . And God said, 'Behold I bring the flood of water on the earth to destroy all flesh which has in it the breath of life, from under the heavens all which is on the earth will perish' " (Gen. 6:6, 17). The Flood was the corrective measure to right a world gone wrong.

The possibility for this dramatic Divine about-face had been kept open. Prior to God's statement of regret, no Divine covenant had been made with humans, no guarantee that we would survive. Though readers of the Bible are informed that Adam and hence all of humankind were created in the "image" of God (Gen. 1:27), this information was only revealed to humankind at the time of Noah and even then only after the Flood (9:6). When God declared the impending destruction of humanity, this was not a rescinding of any previously stated covenant, contract, or intent. No promises of survival had been made with humankind. God kept trying to nudge humans along a morally acceptable path, first by scolding Adam and Eve and expelling them from the Garden, then by exiling Cain following Abel's murder, but all to no avail.

The remedy for our intransigent and defiant nature was the Flood and then reducing our life spans from nine hundred to a hundred or so years, expanding our diets to include meat (Gen. 9:3), and instituting a list of moral and civil guides (9:1–7). With these in place, God established the first covenant with humankind. "And God blessed Noah and his sons . . . saying, 'And behold I establish My covenant with you and with your progeny after you . . . neither will all flesh be cut off any more by the waters of the flood. . . .' And God said, 'This is the token of the covenant . . . I have set My [rain]bow in the cloud that I will remember My covenant between Me and you and every living creature . . . the everlasting covenant' " (9:8–17). Just prior to this God had guaranteed that there would be a consistency in nature: "And God said, . . . 'While the earth remains, seeding time and harvest time, and cold and heat, summer and winter, day and night, will not cease' " (8:21–22).

Why all these Divine guarantees? Did God feel that Noah and family needed encouragement to carry on with life after the colossal Divine punishment of the Flood? Whatever the Divine intent was, it failed in its goal. God clearly underestimated recalcitrant human nature, which we must constantly remind ourselves was created by God.

Just three verses after God completed the Divine rapprochement with Noah and family, Noah planted a vineyard, got drunk on wine, fell into a drunken stupor, and was debauched by one of his three sons (Gen. 9:20–27). That dismal event was topped only five generations later by humanity's rebellion against God in the attempt to build the Tower of Babel (11:1–9). It's no wonder God decided no more floods. God had "learned" that, with human nature being what it is, massive floods are not effective as a teaching device. Anthropomorphically, one can almost see God burying His head in His hands, shaking it back and forth, and wondering if we'll ever learn. Was all this a surprise to God?

Ten generations were to pass after God's conversation with Noah before God again had contact with an individual. It was then that God sent Abraham and Sarah to the land of Canaan, current-day Israel (Gen. 12), and ultimately established a covenant with Abraham (17:1–2). Until Abraham, God had tried a group approach with Adam and Eve's progeny and then the progeny of Noah and family, the biblical equivalent of all humanity. But those attempts failed miserably. With Abraham, God decided a limited venue might be more appropriate considering our human limitations: start with an individual and hope that this influence will eventually extend to all humanity. "And God said to Abraham, 'In you shall all the families of the earth be blessed'" (12:3).

Nahmanides brings from Kabala a remarkable, and at first glance even heretical, explanation of what went so wrong by the time of Noah. But he does so through a two-thousand-year-old parable, that of a king and an architect, telling us that it is one of the deep secrets of the Bible. Just as material physics probes nature to discover the laws embedded therein, so the spiritual physics of kabala seeks to understand, through the episodes in the Bible, the spiritual workings of the Creator within the creation It brought into being.

In the parable, a king wanted to build a palace. But kings don't build palaces. Kings hire architects, and the architects build the palaces. So the king hired an architect. The architect built the palace. The king came to see the palace and was not pleased. At whom should the king be angry? At the architect.

The king wants a palace. The Creator wants a universe. Clearly the king in the parable is the King of kings, the Eternal God. The puzzle is to identify the architect. Could it be nature? Recall that *Elokiim,* the aspect of God made manifest in nature, is the only name used for God during the entire six days of creation. It was during those evocative six days that the universe progressed from being an unformed void (Gen. 1:2) to providing a nurturing home for humanity (1:27). In essence, nature, as guided by *Elokiim,* was the architect of the physical world, and that world included humans with nine-hundred-year life spans.[2]

Of course the Eternal God and *Elokiim* are one and the same, *Elokiim* being one aspect of the all-encompassing Eternal God as made manifest in nature. And as we learned in the previous chapter, nature had been given some leeway, the *tzimtzum* inherent in creation, in how it performs God's wishes.

Once again, we are forced to confront the paradox of a logically faultless Creator finding fault within an aspect of Its own creation. And hence the verse in which we are first told that the Eternal God repented of having made humankind concludes with the anthropomorphism "And it grieved Him in His heart" (Gen. 6:6). In human terms, God seems to be learning what it means to run a universe literally permeated with free will.

The events that follow the Flood reveal a raft of insights into the workings of God in this very human world. When God chastised Cain for being annoyed at the rejection of his sacrifice, God did not provide Cain with any clear guidelines as to how he might improve his behavior. The result was that Cain in jealousy murdered his brother. That was a method of education not to be repeated. Nine generations later, following the Flood, God attempted to intervene more aggressively in the repair of society. Explicit Divine regulations were imposed. A careful reading of Genesis 9 reveals that God decreed seven basic laws incumbent on all humanity. The Talmud in the section dealing with the high court, the Sanhedrin, lists them. Collectively they are known as the seven laws of Noah:

Courts of law are to be established.

Blasphemy is prohibited.

Idolatry is prohibited.

Incest is prohibited.

Murder and physically harming others are prohibited.

Robbery in all its forms is prohibited.

Maiming living animals to eat of their flesh while they live is prohibited.

Of the seven laws, five relate to interpersonal contacts, only two to our relations with the Divine. The rampant robbery and treachery of the generations prior to the Flood could have been controlled by the five directives, and the two others would have sufficed to keep humanity within the range of an ethical theology. Notice that belief in one Creator God is not among the laws. Within the scale of these seven laws, social harmony ranks five to two over our spiritual needs. The Talmud metaphorically quotes God as saying, "Would that they would forget Me and just keep My ways."

The Flood was over and Noah, family, and the accompanying menagerie disembarked the ark. Noah then offered burnt sacrifices to God: "And God smelled the pleasing aroma; and God said in His heart, 'I will not again curse the ground because of humanity for the imagination of a human is bad from youth; and I will never again smite all life as I have done. . . . Neither shall there be again a flood to destroy the earth. . . . The waters shall not again become a flood to destroy all flesh' " (Gen. 8:21; 9:11, 15).

In the verses directly following the Flood, four times God repeats the vow not to bring on another flood. After all, there's nothing to be gained. Humans can't help but be flawed. Their imaginations are "bad from youth." But wasn't God aware of the extent of this limiting human trait prior to the Flood? Had God just learned something new about God's own creation? It would seem so. Again we find *Ehe'ye asher ehe'ye,* "I will be that which I will be," the manifestation of God changing as, in human terms, God learns how to relate to the creation It created. At this point God resolves to make do with the less than perfect version of humanity. The human propensity to err notwithstanding, God commanded Noah and family to repopulate the world (Gen. 9:1).

Violence between neighbors preceded the Flood. Collusion and friendship marked the mood of the populace after the Flood and led them to build the Tower of Babel (Gen. 11:1–9). The Tower was to be a focus, a point of reference able to be seen even at a distance. The goal was to keep the populace in the general locality: "And they said, 'Come let us build a city and a tower with its top in heaven . . . lest we be scattered abroad upon the whole face of the earth'" (11:4). Yet this was contrary to God's explicit request that the people spread out over the entire habitable world (9:1).

Here we find friendship and cooperation, a powerful contrast to the personal violence that led to the Flood. The Divine response to this amicable communal attempt to concentrate the world's population in one location rather than follow God's command to populate the entire world was the dispersion of the populace. This is in sharp contrast to God's reaction to the social violence that induced the Divine decision to destroy the populace via the Flood. The Divine message is that the God of the Bible cares intimately about how we care for each other—even more than how we care about Him.

When viewed from the perspective of a parent-child relationship, God's "emotional" response was quite "human." As any parent knows, when children argue with a parent (the Tower of Babel analogy), it is annoying, but the wise parent's reaction takes into account this rebellion as a logical developmental attempt to assert independence and individuality. However, when my wife and I hear our kids arguing among themselves (as did humankind argue in the time of Noah), even though we are not part of their feud, the painful frustration we feel makes us reach our boiling point. Parents crave joy, not conflict, for their children. As does God for His creation. Of course no sooner have I boiled over, as I unsuccessfully attempt to quell the quarrel, than I regret my having entered their fray, usually at least four times over.

When God tells us four times no more floods, I can empathize with this Divine evidence of "frustration." At times God sounds almost human. But then, as the ancient sages tell us time and again, the Bible talks in the language of humankind. The very fact that the Bible by necessity uses human language lowers it to our mundane level. There is, however, no option other than language for explicit communication.

To err, as the philosopher Hannah Shir points out, is part of life's learning process. We study history in an attempt to learn from the successes and errors of the past. From the biblical account of the Flood and its aftermath, it seems that there is room in our created universe for this learning process within the Divine as well.

Because of our misdirected choices in the generation of Noah, God regretted having allowed the world to develop as it did and chose a new path for its continuation. God realized the world needed to be rewired and acted on that new Divine realization. As the Bible makes clear in incident after incident, this is a God that can only be described by the dynamics of a verb reaching into the future. How could anyone have thought the static "I am" could describe such a Force?

SIX

Arguing with God

We Are Partners with God and Partners Can Disagree

Though God evicted Adam and Eve from the Garden because of their rebellion and later strictly punished the wayward people with the Flood and also at the Tower of Babel, these are examples of extreme defiance of God's intent. In other biblical episodes the protagonists actually bargain with God over the extent to which God's demands need be fulfilled. Rather than expressing absolute refusal to comply, people can make use of Divine channels that are open to argument. The most classic biblical case of this right to protest is found in Abraham's relationship with God and with Sarah, his wife. This couple founded the Hebraic religion.

Abraham is called "the Hebrew" at the time of the capture of his nephew Lot by pagan kings (Gen. 14:13), and Abraham's descendants enslaved in Egypt are referred to as "the Hebrews" by God just prior to the Exodus (Exod. 3:18). The three letters of the Hebrew root of the word "Hebrew," *ayin, veit, reish,* spell *aver,* which means "over or from the other side," as in "[And God said,] I took your father from the other side of the river" (Josh. 24:3). Being from "the other side," Abraham was very different from his contemporaries. Abraham and his wife, Sarah, had rediscovered the Eternal God of monotheism. They disseminated among their contemporaries this empowering realization, the teaching being referred to in the Bible as "they had made souls" (Gen. 12:5). Having chosen God, God chose them.

Abraham had discovered God even though he had been raised among idolaters, in a culture where the deities took on material form. But all was not smooth in Abraham and Sarah's marriage. For decades, Sarah was unable to conceive. It was not until Sarah was eighty-nine years old and Abraham ninety-nine that the Divine blessing came, and they conceived what was to be Sarah's only child. When Abraham, assuming that he and Sarah were far beyond the child-bearing age, laughed at the news of the impending birth, God told him to name the baby Isaac, which in Hebrew (*Yitz'hak*) means "I will laugh" (Gen. 17:17, 19).

God informed Abraham that Isaac was to carry on the Divine covenant that God had made with Abraham and Sarah. But then, to our surprise and consternation, some decades after Isaac's birth we are confronted with a most incongruous demand. God directs Abraham to sacrifice Isaac: "And [God] said, 'Take your son, your only son whom you love, Isaac, and get yourself to the land of Moriah and offer him there for a burnt offering on one of the mountains that I will tell you of' " (Gen. 22:2). We can only imagine the confusion Abraham felt. And yet Abraham willingly carried out this Divine command, and did so with alacrity, as indicated by the wording of the following verse: "And Abraham rose up early in the morning and saddled his donkey and cut the wood [to be used for Isaac's funeral pyre]." Abraham wanted to fulfill every aspect of the task by himself, not have any part of it done by a servant.

A Talmudic legend claims that the emotional test of the sacrifice of Isaac was in recompense for an earlier act by Abraham and Sarah against Abraham's first child, Ishmael. For the first ten years of marriage, Sarah had been unable to conceive. At that stage, Sarah urged Abraham to also marry her maid, Hagar, and to build their family through Hagar's progeny. Such an arrangement was common at the time. The result of that union was Ishmael, progenitor of the tribes of Arabia and eventually the religion of Islam. Isaac was born when Ishmael was thirteen years old.

Ishmael, perhaps jealous at having been displaced as the only child, "made fun" of Isaac (Gen. 21:9). As a result, Sarah was unwilling to allow Ishmael to remain in the household. She ordered Abraham to banish both Hagar and Ishmael. The expulsion pained Abraham deeply, "was very grievous in Abraham's eyes" (21:11), but at God's direct command, Abraham submitted to Sarah's demand. Abraham sent Hagar and Ishmael away with a jug of water and a loaf of bread (21:14). Abraham was wealthy, having flocks of sheep and bands of servants. The legend asks why Abraham sent Hagar, his wife, and Ishmael, his child, off with just water and bread. Why not with herds of cattle and servants and gold and silver? Abraham's "punishment" for treating Hagar and Ishmael so shabbily was the "binding of Isaac," as the near slaughter of Isaac is referred to in Judaic literature. A later biblical law teaches that if a man has two wives, even if the more loved wife gives birth second and the less loved wife births first, the firstborn child must receive the main inheritance (Deut. 21:15–17).

Abraham's compliance with this Divinely ordained sacrifice of his son Isaac represents a pivotal episode, perhaps the pivotal episode, in the Hebraic religion. It is viewed as the ultimate measure of Abraham's devotion to God. Yet considering that the eventual conquest of Canaan by Abraham's descendants was justified in God's balance by the abominable Canaanite practice of child sacrifice, the summoned sacrifice of Isaac seems totally inappropriate. In fact, it seems the complete opposite of the values later espoused by God: "[And God said,] 'For every abomination . . . they [the Canaanites] have done for their gods. Even their sons and daughters they burn in fire to their gods'" (Deut. 12:31). Lest that verse appear to be the biblical rewriting of history to justify the conquest of Canaan, excavations in modern-day Israel have revealed altars from the Canaanite period surrounded by "cemeteries" containing multitudes of scorched infant skeletons.

Abraham was raised among those who worshiped idols. Child sacrifice was the order of the day. God's command to slaughter Isaac was completely in keeping with the local customs of Abraham's youth. Abraham was seventy-five years old when God told him to leave the idolatrous surroundings into which he had been born. He was a hundred years old at Isaac's birth, and by the time of God's command, Isaac was already old enough to carry the wood intended for his pyre (Gen. 22:6). Over three decades had passed since Abraham left the idolatrous Mesopotamia of his youth. He'd had intimate contact with God while God guided his exodus from Mesopotamia into Canaan (Gen. 12). At this advanced age he must have realized that the Eternal God whom he and Sarah had rediscovered was a God of life, not of death.

Even assuming that God had some inscrutable plan, the question arises: Why didn't Abraham argue with God? Abraham might have reasoned with God that the command to kill Isaac ran counter to the very nature of the Divine theology the Eternal God represented. If God rejected the contest and continued to demand the sacrifice, then Abraham could have complied.

Long before the command to sacrifice his son, Abraham had argued with unrelenting force when God informed him of the impending destruction of the wicked cities of Sodom and Gomorrah. For ten biblical verses Abraham "stands before God" and demands, "Will You indeed sweep the righteous with the wicked? Perhaps there will be fifty righteous in the city, . . . forty-five righteous, . . . forty, . . . thirty, . . . twenty, . . . ten . . ." In that debate, Abraham acknowledges that he is "but dust and ashes," yet Abraham reminds God: "Forbid that You would do a thing as that. . . . Shall not the Judge of all the earth do justly?" (Gen. 18:23–32).

A traditional reply to the puzzle of Abraham's silence at the binding of Isaac might be found in the customs of the day. Abraham may have believed that child sacrifice was the ultimate pledge of allegiance to God. Certainly that was the way of his neighbors. When the angel of God stopped the sacrifice at the last moment, the message was that allegiance to the God of the Bible stops at murder. But first Abraham had to personally realize that his allegiance to God was as great as that of his idolatrous neighbors to their own gods. In this sense, the experience was to prove to Abraham, not to God, the extent of Abraham's belief in and loyalty to God.

This rather docile acquiescence by Abraham at the command for the sacrifice stands in sharp contrast with the ultimate name, Israel, given by God to Abraham's progeny. The renaming came from an experience when Jacob, the son of Isaac, actually wrestled with a messenger angel of God. The background to this wrestling match sets the stage for the renaming of the clan. Jacob had gone to Mesopotamia, the homeland of his mother, Rebekah, in part to escape the wrath of his brother, Esau, who felt Jacob had cheated him out of his rightful eventual claim to the family inheritance, and in part to seek a wife "from the old country." In

Rebekah's opinion, the local Canaanite women were below par spiritually. In the ensuing marriage contracts, Jacob was obliged to remain in Mesopotamia for two decades. Upon his return to Canaan, Jacob received word that his quite bellicose brother, accompanied by four hundred men, was coming to meet him. Jacob feared that a military confrontation was in the offing. He divided his camp into several sections as a strategy that might allow partial survival if the battle did transpire.

And then at nightfall he went to retrieve some forgotten items that had been left a distance for the main camp. Alone in the dark of night, he was attacked by a stranger: "And Jacob was left alone; and there wrestled with him a 'man' till dawn. And when he [the man] saw that he did not prevail, he touched the hollow of his [Jacob's] thigh. . . . And he [Jacob] said, 'Bless me.' And he said, 'What is your name?' And he said, 'Jacob.' And he said, 'Your name will no longer be called Jacob, but *Israel,* for you have striven with God and with men and you have prevailed' " (Gen. 32:24–29). The name Israel derives from the Hebrew roots *yisrah,* "to prevail" or "to overcome," and *ail,* the aspect of the Divine manifest as power. The combination of the two roots yields *Yisrael,* meaning a people who "wrestle with and strive for God."

Jacob wrestled, and Jacob prevailed. And Jacob was praised for this persistence by the "man" with whom he wrestled. God wants us to wrestle, wants an interactive relationship, a dialogue, a sharing of power and responsibility between the Creator and the created. Just as the manifestation of God's nature is dynamic ("I will be that which I will be"), so God wants a dynamic relationship with humankind. Just as a couple builds a relationship through dialogue, so God wants our input. And yet Abraham voiced not the slightest hint of a protest.

Isaac had done no evil, at least none recorded in the biblical text. There was no human moral justification for his sacrifice.

If the impending slaughter of Isaac pained Abraham, there is no mention of that emotion, only an almost robotlike compliance. The mitigating factor in the debate over the intended sacrifice might have been that in the argument to save the righteous of Sodom justice was at stake. Why should the innocent of that city suffer along with the guilty? The sacrifice of Isaac represented a test not only of justice, but also of allegiance to God, returning to God the gift of this lad, Isaac, whom through a miracle God had given to Abraham and Sarah in their old age. Yet what Divine purpose or eternal lesson could there be in killing an innocent child?

During the Exodus, God made it exceedingly clear that God wants an argument when the Divine plans seem humanly unjust. Barely two months had passed since the redemption of the Israelites from slavery in Egypt, when the Israelites rebelled against God. God decided the people must be destroyed. But before acting, God informed Moses of His plans. "And God said to Moses, . . . 'Let Me alone that My anger may act against them and I will consume them' " (Exod 32:9). "Let Me alone"? God doesn't need Moses's consent to act. So why the "Let Me alone"? God has all but begged Moses to argue for the Israelites' salvation. And argue Moses does. "[And Moses said,] 'Wherefore the Egyptians will say, "For evil He brought them out to kill them in the mountains. . . ." Turn from Your fierce anger and repent of the evil against Your people' " (32:12). "And God repented of the evil which He said He would do to His people" (32:14).[1]

The God of the Bible is full of surprises. Here God literally sets up Moses to act as the defense attorney for the people of Israel, God's "let me alone" request, and then has Moses defeat His charge. This is a God who not only has regrets, but also is willing to accept a good argument and then to repent when the argument successfully refutes His proposed plan.

Abraham was not one to acquiesce passively to the challenges he confronted. When his nephew, Lot, was kidnapped during a battle between neighboring "kings," Abraham immediately led his troops to the rescue (Gen. 14). This proactive aspect of Abraham's character makes his silence at the requested sacrifice of his innocent child all the more puzzling.

It is obvious from the outcome of the binding episode that God never intended to allow Isaac to be slaughtered. The question is why Abraham didn't realize this and act accordingly, that is, argue. There was ample opportunity. The Bible informs us that three days passed between God's command and Abraham and Isaac's arrival at the site for the sacrifice.

What was God's "test" of Abraham (Gen. 22:1)? Was it Abraham's faith in God, or was it Abraham's knowing that God was a God of life and not of death? Abraham passed the test—that of obeying God, and so he and his progeny were to be rewarded. However, based on the narrative that follows the attempted slaughter, Abraham passed the trial differently from the way God had originally envisioned it.

I submit that, indeed, God had expected an argument and, furthermore, that the absence of a challenge by Abraham disappointed God.[2] Prior to the binding of Isaac, God had frequent contact with Abraham—promised him progeny with a great future, acknowledged that He knew Abraham in a way more deeply than He knew others (Gen. 18:19). After the binding, direct contact between God and Abraham totally ceases. To emphasize this, at the last moment before the slaughter, when Abraham is told, "Lay not your hand on the boy and do nothing to him" (22:12), the Eternal God does not bring the message. Rather, the command is relegated to an emissary, an angel: "And an angel of the Eternal God called to him from heaven and said, 'Abraham, Abraham.' And he said, 'Here I am.' And [the angel] said, 'Lay not your hand on the boy'" (22:11–12). Recall that following the episode in which Abraham argued with the Eternal God prior to the destruction of Sodom, the dialogue between Abraham and God continued. Only when Abraham failed to challenge God for the seemingly unjust demand of his child's sacrifice did God diminish the intimacy of their relationship.

It is true that following Abraham and Isaac's submission to the binding and its abortion, Abraham is promised that "Because you have done this thing and not withheld your son . . . I will bless you and . . . I will multiply your seed as the stars of the heavens and as the sand upon the shore of the sea" (Gen. 22:16–17). Abraham hadn't failed the test. He had complied with God's demand and accordingly Abraham is indeed blessed for his compliance. But again, who brings the news of this grand blessing? It is not a direct transmission from God. The blessing is conveyed to him by an intermediate, an angel.

If you have been used to getting phone calls directly and personally from the president and suddenly the presidential messages come through a secretary, you notice the change immediately. A parallel to this lowering of profile, but on a human level, occurs generations later in Egypt with Joseph, son of Jacob and great-grandson of Abraham and Sarah. As a consequence of Joseph's mapping out a plan to save that vast country from starvation during the impending seven years of famine, he was made prime minister. All the time he was actively involved in the national rescue of Egypt, he had direct access to Pharaoh. But less than two decades later, the crisis being over, Joseph was no longer crucial to Egypt's survival. When his father, Jacob, died, Joseph asked permission to bury Jacob in the Hebron cave of the patriarchs: "And Joseph spoke to the house of Pharaoh saying, 'If now I have found favor in your eyes, please speak in the ears of Pharaoh saying . . .'" (Gen. 50:4).

The wording makes it very clear. Joseph all too well realized the difference between the one-on-one contact to which he had become accustomed in his earlier years and the level to which he had now been reduced, begging contact by way of an intermediary, speaking to "the house of Pharaoh," ingratiating himself ("If now I have found favor . . .") and asking others to speak for him. How much more so must Abraham have felt the change when relating to his Creator.

When Abraham and Isaac ascended the mountain, "They went the two of them together" (Gen. 22:6). When Abraham descended, after the binding, Abraham was alone (22:19). The next event recorded in the Bible is the death of Sarah (Gen. 23). Did the news of the attempted sacrifice steal her soul? If so, then Sarah, not Isaac, became the casualty of the binding. As Rashi points out in his commentary, fright can kill. Though the burial of Sarah is described in great detail, Isaac is not mentioned as being present at Sarah's burial. Yet we learn later of Isaac's deep attachment to his mother. It was not until he married that the gap left in his life by her death was filled. Only then was Isaac "comforted after [the loss of] his mother" (24:67).

Could being "comforted after [his loss of] his mother" relate to an unspoken lingering fear held within Isaac's heart that the attempted sacrifice caused his mother's death? Does Isaac feel he should have protested, even though his father was willing? Perhaps Isaac did protest. Recall that he was tied, bound, on the altar. Was he old enough to resist? The Bible does not inform us of Isaac's age at the time of this incident. However, the biblical chapter immediately following records Sarah's death: "And the life of Sarah was one hundred twenty-seven years; these were the years of the life of Sarah" (Gen. 23:1). Sarah gave birth to Isaac when she was ninety years old (17:17). If her death followed closely the near miss of the binding, then Isaac would have been thirty-seven years old. Though the English word used to describe Isaac at the time is "boy" or "lad," the Hebrew is *na'ar,* meaning a male of marriageable age. At thirty-seven, Isaac likely could have resisted. However, in the strong paternal society of the time, the powerful tradition of strict filial obedience undoubtedly would have inhibited such a response, even if death hung in the balance.

There is no indication that Abraham had told Sarah of God's request, no recorded dialogue, though there was considerable debate when Sarah demanded that Abraham exile Ishmael, Abraham and Hagar's son, from the household shortly after the birth of Isaac. If Isaac had perished at Abraham's hand, then having banished Ishmael would have been to no avail. The only surviving progeny of Abraham would have been Ishmael. Was Isaac absent from his mother's burial to avoid an unpleasant encounter with his father?

The binding seems to have caused a family trauma. Abraham was in Beersheba at the time of Sarah's death (Gen. 22:19). Sarah at the time of her death was in Hebron (23:2). Isaac lived near Be'er-lahai-roi, "the well of the Living One that sees" (24:62). Interestingly, after the ordeal of the binding, Isaac chose to live in the same locale to which Hagar fled when Sarah chastised her, Be'er-lahai-roi (16:14).

Isaac had no further contact with his father during Abraham's life, though Abraham made the successful effort to find Isaac a good wife (Gen. 24). But note how Abraham chose this wife. Without any hint of having consulted Isaac, Abraham sent his servant to Abraham's former homeland to seek a proper bride for Isaac. Yet Isaac was not a youngster by this time, so why not send Isaac himself? If indeed Isaac was living with Ishmael, then Abraham must have been concerned about the questionable influence of Ishmael upon Isaac's choice of a bride. Hence Abraham's affirmative action to find a spouse for his and Sarah's son.

But again, why not send Isaac? Couldn't he be trusted to make a good choice? There had been no Divine demand that Isaac not leave Canaan. A generation later, Isaac and his wife, Rebekah, have no problem in sending their son Jacob back to that same country, partly to escape brother Esau's wrath, but also to allow Jacob to choose a proper wife (Gen. 28:1–2). Both Jacob's and Isaac's future wives enter the text in the same setting, at a well watering their fathers' flocks of sheep. Isaac accepts his prechosen wife-to-be without a protest, even though he is introduced to her via his father's servant and not by Abraham himself. Though he may have been estranged from his father, the culture of the day demanded that he accept the paternal choice of a wife, just as he had accepted his father's willingness to "bind" him.

Following the aborted sacrifice, the only recorded contact Isaac had with his father was after Abraham's death. Isaac and Ishmael bury their father in the double cave in Hebron. Abraham had purchased that cave for Sarah's burial. Abraham was buried next to Sarah, mother of Isaac, but not next to Hagar, Abraham's second wife and the mother of Ishmael (Gen. 25:10). Since Abraham had wed both women, this burial arrangement must have exacerbated any enmity that might have already existed between Isaac, the progenitor of the people of Israel, and Ishmael, progenitor of the Arabian peoples, and quite possibly between the progeny of these half brothers. Perhaps Hagar was still alive. When God had appeared to Hagar at Be'er-lahai-roi, God told her that her son Ishmael would grow to be a powerful and wild adversary "in the face of his brethren" (16:12). The prediction seems to have been fulfilled, unfortunately.

The sequence of events at and following the binding give compelling force to the supposition that the God of the Bible not only wants a dialogue with us humans, but even more than that. God expects such, and if the situation seems unjust or unjustified, then, beyond a dialogue, God wants us to argue. If our case is strong enough, God will even "give in," or at least modify the Divine directive. Moses seems to have understood this trait of the Divine.

Six generations after Abraham, when God confronted Moses at the burning bush and told him to return to Egypt to lead the people of Israel out of Egyptian slavery, Moses tenaciously argued with God against the wisdom of God's proposal (Exod. 3–4). God indulged Moses, listened to his pleas, and even related to them. As a parent would respond to a youngster, God modified some details of His command, accepting that Aaron, Moses's brother, would share the task with him. Those changes being in place, God then insisted that Moses comply. And indeed the faithful Moses obeyed. Could Abraham have acted in a similar manner?

Argument seems to be the standard and the *expected* biblical operating procedure in our encounters with the Divine. The surprise is that, having designed and created our universe with all its magnificence and granted us the freedom of choice, God wants us, expects us, to interact with the Divine about how to run the universe.

Bargaining with God likely requires an established "working relationship" prior to the confrontation. At the burning bush, Moses had already proven his moral worth by standing up for the oppressed, whether they were Hebrew (Exod. 2:11–13) or gentile (2:16–20). Moses had a track record upon which God could rely, as did Abraham prior to the incident at Sodom. Abraham and Moses were well known to God before they entered the Divine fray.

The estrangement of God from Abraham following the binding of Isaac finds parallel in the episode of the Garden of Eden and the forbidden fruit. Adam and Eve were exiled from the

Garden for having transgressed God's command and eaten from the Tree of the Knowledge of Good and Evil. Subsequent to their chastisement by God, Adam and Eve have no further contact with God, not even after the tragedy of their son Cain murdering his brother, Abel. This may explain the meaning of God's warning to Adam: "And the Eternal God commanded Adam saying, 'Of every tree of the Garden you may surely eat. But of the Tree of the Knowledge of Good and Evil you shall not eat from it, for in the day you eat from it you shall surely die'" (Gen. 2:16–17).

Following their error, in which they rebelled against God's command, they live for many centuries, reaching 930 years (Gen. 5:5). So how do we account for the Divine statement that in the day of transgression "you shall surely die"? In the Hebrew text, the verb "die" is doubled, hence the "surely die," and yet they lived. They lived, but without the intimate relationship with God that they had previously enjoyed. As a result of their disobedience, they were exiled not only from the Garden. Far more significantly, they were sent out from God's presence. And that exile was death, spiritual even though not physical. Throughout all those centuries of exile from Eden, God never again addresses them.

The Bible does not imply that eating of the forbidden fruit brought physical death for the first time into the world. The death that this first of *human* couples experienced was the death of their unbounded spirituality. Loss of spirituality for one who had conversed with the Creator, a separation from that infinite light, would be far more devastating than actual physical death. For this unfortunate couple of the Bible, only the physical remained.

Cain suffered similarly. At Cain's exile, following his murdering Abel, he pleaded: "My punishment is greater than I can bear. . . . From Your Presence I shall be hidden" (Gen. 4:13–14). Cain, who had known God, talked with God, was now consigned to an existence totally confined to the material. He chose to murder, and with that choice he lost the spiritual dimension of life, which, as he acknowledged, was the most severe of punishments.

At the Divine rescue of Isaac the exact wording of the verses is instructive: "[And the angel said,] 'Do not send your hand upon the boy, neither do anything to him for now I know that you fear God [or 'are in awe of God'; the Hebrew word *irah* can mean either 'fear' or 'awe'] and have not withheld your son, your only son, from Me" (Gen. 22:12). The angel acknowledged Abraham's fear and awe of God. The realization that the Creator can take life and then in a flash restore it is awesome. And child sacrifice is bound to induce fear. It would, however, seem hard to love such a God. And that being the case, the sacrifice was aborted at the will of the Eternal God.

It is taught in Kabala that *irah* ("fear," "awe") can lead to love, *a'ha'va*. That may be true. However, love is the biblical goal, both in our approach to the Divine and in relationships among humans. Two verses connect the central concept of biblical religion with love. "Hear, Israel, the Eternal our God, the Eternal is One" (Deut. 6:4; quoted in Mark 12:29) is followed immediately by, "And you shall *love* the Eternal your God with all your heart, with all your soul, and with all your might" (Deut. 6:5; quoted in Mark 12:30; Matt. 22:37). The primary Divine objective is to bring love, not fear or even awe, into the world.

This discussion of the binding of Isaac and its aftermath is not intended to teach us how Abraham, the founder of the people of Israel, "should have" acted. In his time child sacrifice was what was done, although Sarah, it seems, might have thought otherwise, had she been consulted. Rather, the episode brings the message of what God wants of us, how we are to act and react when challenged by life's vicissitudes. We have the right, in fact the Divinely granted duty, to dissent when life presents us with demands deemed unjust and undeserved. Anything less than that betrays a misunderstanding on our part of God's interactive role with Its creation. In proportion to the relationship that we have established with God during times of joy, we can demand Divine redress in times of trouble. As we would do in any loving relationship, we can argue with God. That in itself can lessen the burden.

SEVEN

In Defense of God

Understanding God Through the Book of Job

We cannot complete an exploration of the character of God on display in the Bible without looking at the book of Job. In the annals of undeserved suffering, just a sliver of pain below the million innocent children murdered by the Nazis and the bug-infested orphans of Botswana comes the biblical character of Job.

Job was wealthy and kind, a family man who dearly loved his children, one who we surmise from the text cared for the poor and needy. In brief, life was good to Job, and Job was good to life. Then with no warning catastrophe struck. Blow upon blow beat down upon him. First his material possessions were destroyed by acts of nature and stolen by marauding gangs. Then Job's children were killed by a violent wind that crushed their house upon them. Job mourned his losses, but stoutly retained his faith in a just and loving God: "Naked I came from my mother's womb and naked shall I return. The Lord gave and the Lord has taken" (1:21). With his children and material wealth gone, Job was then stricken with incurable sores. Lesions and boils covered his body, so that agony accompanied his any and every motion. His wife, seeing Job's torment, urged him to curse God and die. But this is not the Job of the Bible. "Shall we receive the good from God and not also the evil?" he asks (2:10).

When Job's friends learn of his plight, they come to offer consolation. Job laments what he perceives as his undeserved fate and professes his innocence. At this point his visitors do anything but console. Instead, they offer what was then, and is to a large extent now, the standard pseudomoralistic take on life—you get what you deserve. If you are suffering, then search your past to discover your transgressions and repent of them. Since God is righteous, you must be wanting. And of course, when we take these events as a whole, Job's murdered children must also have been far less than righteous, or how could we justify their untimely deaths? And his wife also, now plunged from affluence into abject poverty. A good, all-powerful, all-loving, all-knowing God makes no error in judgment, allows for no unwarranted suffering, or so these friends of Job profess.

But we know the total untruth of this all too common mantra, because we have read the opening passages of the book of Job and therefore know something that obviously neither Job nor his friends could have known. We've looked behind the screen of the material world and been privy to information of which Job and his solicitous friends are totally unaware.

The Eternal God had openly lauded the faithfulness of Job. When God sang Job's praises, the Adversary, named Satan in the book of Job, countered, claiming Job's entire allegiance to God was the result of Job's being blessed with a loving family and immense wealth. "Put forth Your hand," said Satan to God, "and strike all that he has. Will he not then curse You to Your face?"

And the Eternal said to Satan, "Behold, all that he has is in your hand" (1:11–12).

We know that the entire story of Job's grief that is about to unfold results solely from a wager between God and Satan, a wager to determine whether Job will abandon his faith in God in the face of his affliction. We are never told just who or what Satan actually is supposed to be or represent—perhaps he is a fallen angel or some fantasy of our inclinations—or, more bewildering, how such a "being" could induce God to take a wager that would bring the grief of death and destruction to a faithful devotee.

Maimonides, in his classic *The Guide of the Perplexed,* attributes the root of the word Satan to the Hebrew *s'teh,* "turning away," a different form of the word that is used in Proverbs 4:15: "Avoid it; do not pass by it; turn away (*s'tai*) from it." The goal of Satan is to lead his target from the path of truth. The concept of Satan as a force actively attempting to change our path is already found in Numbers, the fourth book of the Bible. In the account of Bilaam and his donkey, we read, "And God's wrath was kindled because he [Bilaam] went; and the angel of God placed himself in the path 'to *satan* him' " (22:22). Here, the literal Hebrew for "to *satan* him," though often translated "for an adversary," reads "to divert him."

We've learned from the book of Job, and that means we've learned from the Bible, something extraordinary. God does test His loved ones with unearned suffering in order to show their worth. But to show it to whom? To Satan? And what test was there for Job's murdered children, his impoverished wife? Was God's command for Abraham to bind and sacrifice Isaac on a fiery altar a similar test (Gen. 22)? We only know the information recorded in Genesis. Is there a similar tale of Divine praise and then a hidden, behind-the-scene wager between God and the

Accuser that underlies Abraham and Isaac's trial? The Talmud (*Sanhedrin* 89b) suggests that such indeed was the case.

The Talmudic passage opens with a telling quote from Genesis:

> "And it came to pass after these things that God tested Abraham" (Genesis 22:1). After what things? Rabbi Yohanon, quoting the sage Rabbi Yosi Ben Zeimra, said: "'After these things' refers to the words of Satan, as it is written, 'And the child grew and was weaned and Abraham made a great feast on the day Isaac was weaned' (Genesis 21:8). Satan said in front of The Holy One Blessed Be He, 'Ruler of the world, this old man was given a son at one hundred years age. In all his feasts did he ever offer to you as a thanksgiving sacrifice even a turtledove or a pigeon?' [The Holy One Blessed Be He] replied: 'The feasts are wholly for his son. If I were to say to him: sacrifice your son, immediately he would do it.' Immediately after these words, comes 'God tested Abraham'; and it is written 'Please take now your son . . . and offer him as a sacrifice.'"

As we say, the rest is history, or at least biblical history.

In the opening of the book of Job, God extols the virtues of Job. Similarly, in Genesis, God praises Abraham's faith and worthiness and claims to have special "knowledge" of him: "Abraham shall surely become a great nation, and all the nations of the earth shall be blessed in him. For I have known him . . ." (Gen. 18:18–19). Nahmanides asks, "Doesn't God know everyone? Why do I need the Bible to tell me that God knows Abraham?" We are told this because God's knowledge of Abraham was special, more intense than God's knowledge of other persons. According to the Talmud, this Divine praise of Abraham was a springboard for a challenge by Satan, the Adversary, to have God test Abraham. The Talmud tells us that there were earlier tests, but those were less challenging. Ordering Abraham to offer Isaac as a sacrifice was the ultimate. Indeed, it was. From the text of Genesis we can never know if this confrontation between God and Satan actually happened. But without it, God's challenge to Abraham to sacrifice his son seems a bit extreme.

Abraham passed the test. But if Sarah's death was the result of shock upon learning the news of the "near miss," the almost slaughter of her son Isaac, as Rashi suggests, then the real victim of this trial was Sarah. As noted in the previous chapter, biblically we are informed of Sarah's death immediately following the binding of Isaac (Gen. 22:1–19; 23:1–2).

There are intriguing parallels between the account of Job and his sufferings and Abraham's offering of Isaac as a Divinely commanded sacrifice. Following the calamities that struck Job, Job gets a new family. Following the binding of Isaac and the ensuing death of Sarah, Abraham remarries and has many children by his new wife (Gen. 25:1–6).

There is, however, a significant difference in how God relates to Job and Abraham in the aftermath of their trials. Following Job's misfortunes, God speaks to Job directly. After the binding of Isaac, God never again addresses Abraham. All the future Divine communications with Abraham are indirect, by means of an agent, an angel. Could that difference in God's subsequent relationship to Job and Abraham be the result of the difference in how these two related to God during their respective trials? Does God relate to us as we relate to God? Job saw the total injustice of a supposedly loving God dragging him through such grief, and Job argued strongly that he deserved a lot in life far better than what was being dealt to him. Abraham apparently assumed that the way of God was to ask for child sacrifice, which is more than a bit insulting to the Divine image as projected into our finite world.

The closing of the book of Job is as confusing as the opening wager is incongruous. God allows Job to build a new family (not rebuild, since the former family members are all dead) and to regain his wealth. When God in a whirlwind and storm finally addresses Job, God never informs Job of why he suffered tragedy. Nor does God in any way hint that Job, through wrongdoings, deserved the punishment he had received. Nor does God "apologize" for the inflicted trauma. Job is, however, told in no uncertain terms that neither he nor any other human can understand the complex, intricate workings of the universe. Human ingenuity and intelligence may be able to design computers and send people into space, but there is a limit to our knowledge, and beyond that limit lie the ways of God. Isaiah, whose prophecy during the fall of the northern kingdom of Israel, 721 BCE, encouraged his people not to despair, puts it very succinctly: "For My thoughts are not your thoughts; neither are your ways My ways, says the Eternal God. For as the heavens are higher than the earth, so are My ways higher than your ways and My thoughts than your thoughts" (55:8–9). In other words, "I will be that which I will be." You'll have to settle for that.

God, having confirmed that Job is innocent of evildoing, chides Job's friends for having implied that Job was less than righteous. Yet, his virtue notwithstanding, Job suffered greatly. For Job to have maintained his love of God, as he did, Job broke with the widely held, erroneous concept that in a universe created by a loving and powerful God, God will protect His righteous ones from harm. This leads to what may be the most significant and far-reaching message of the book. Even though Job-like tragedy occurs, whether brought on by acts of nature or human ill will and error, still God is present, knowing and, most important, caring. We see this in the fact that God, though not directly interfering to terminate the suffering during the trials, limited Satan in how severely Satan was allowed to trouble Job: "And God said to Satan, 'Behold he [Job] is in your hand, but only be sure to guard his life'" (1:12). And eventually God exiled Satan from the picture.

God has not designed a mechanical "vending machine" world where you put in two good deeds, pull the lever, and out pops the commensurate comfort of a material reward. What does "come out" is what the deeds themselves put in, the pleasure of a life of purpose and plan. Pleasure and comfort are not partners. Avoiding discomfort can cheat a person of the greatest pleasures life has to offer. Just ask any mountain climber at the peak or parent of a first grader.

We sometimes say that love and hate are opposites. But that is not true. The opposite of love is indifference. With hate there is still strong emotional attachment that often has at its roots what was once, and could be again, love. Indifference evidences the total loss of love's potential. It is the emptiest of relationships. In fact, it is the absence of any relationship. From the incidents we've studied, Abraham's and Job's tests, God's demand for us to argue, God's annoyance at the earth's rebellion in not producing fruit trees that also bear fruit—from point after point we learn that indifference is one powerful aspect that is *not* part of God's relationship and partnership with the world. From the "behavior" of the rocks and water of the earth to the choices made by the mind of a prophet, God is interested and involved in everything. How God manifests love for Its creation may not match how we would envision or prefer such love to be bestowed. But as God told Job, much in this world is beyond human comprehension.

When God established the covenant with Abraham, Abraham was told that his progeny would receive great reward, but that reward would be granted only after four centuries of exile and toil (Gen. 15:13–14). How does one justify to the populace of those four centuries of grief that eventually the sun will shine, but not in their time? And especially how does one do this when Abraham himself is immediately promised that he personally would not suffer the ensuing toil ("But you shall go to your fathers in peace and be buried in a good old age," 15:15)?

Does it really matter to parents as they bury their murdered child that in a few centuries relief will come? They would long for redress in their own time. Abraham could accept the package of a harsh exile followed generations later by reward only because he realized that the entire sequence, from covenant to exile and slavery to Exodus, liberation, and reward, is all part of a single continuum, an integrated whole. Abraham's encounter at the covenant (Gen. 15:17–18) took him behind the veil of our temporal world. Abraham realized the total Oneness of existence, a single fabric woven of a helix stretching out in the flow of time and in the breadth of space.

We find for Abraham's experience a parallel in Job's suffering and future reward. A myriad of individual threads, interwoven through time and across space, combine to form the fabric we and all generations experience as existence. "The Eternal is One" proclaims a unity that not only encompasses the totality of each moment's existence, but also plaits the past and the present with the future. The ongoing community is a single organism, the Jungian communal historical consciousness, which, if astronomy, cosmology, and the Bible are correct, stretches back to the very creation. Awareness of this unity can provide the mind-set that enables us to endure travail as well as to celebrate joy with the clarity that beyond the veil of our material world, with all its limitations and inconsistencies, there lies the reality of the all-encompassing One.

The fantasy God imagined in our childhoods, an all-protecting fatherlike deity who always rewards me when I am doing okay, is a most comforting concept of the Divine. The problem is that that God is not the God of the Bible. From a casual reading of a day's newspapers, it is also not the God of this world. From the Bible's description of the world and from the book of Job, we learn that whether one's lot in life seems fair is irrelevant. The continuing demand is that we be just in our actions. People do not choose the physical or social environment into which they are born, but how they react to that environment is to a very large extent a personal choice. Biblically, the burden of choice rests on the individual, notwithstanding the fact that the reward comes through the community.

As I read the events of the Bible, in human terms I see God in a sort of emotional bind. God desperately wants us to choose life, a dynamic, purposeful existence, but doesn't want to force us along that line. Hence we are granted the liberating *tzimtzum* of creation. God has to hold back and let us try. When we really mess up, God steps in. It's so human. Mom teaches junior to play chess. Looking over his shoulder as her son makes his moves on the board, she sees a trap developing. He is about to lose his queen. If she wants her kid to learn to think ahead, to envision the distant outcome of the initial move before that move is made, she will do well to keep her hands in her pockets and let him make the error or at most give a few very general suggestions, as God through the Bible gives to us. It's frustrating, even painful, but it is part of the learning process, Divine as well as human.

The choice for Job was extreme. Following the tragedies that were inflicted upon him, Job had to choose between remaining a victim, being stuck in the meanness of the past, or becoming a survivor, focusing on what can be built from what remains. Job chose to be the survivor. With that proactive outlook, he chose life and found the love to build a family anew. It was his personal contribution toward the communal obligation to shape a better world, *te'kun olam,* 'to repair a less than perfect world.'

It is not a fantasy that cancer steals the young as well as the old from the pleasures of this world, that earthquakes crush the innocent as well as the less than innocent. Following any tragedy, we can consider ourselves victims and spend the days allotted to us in vengeful anger, mourning the outrages of the past. Or we can realize that we are actually survivors. I have had the privilege to number among my acquaintances people whose families and fortunes were torn away by the Holocaust, who watched their infant children and loving wives dragged to the gas chambers. And then following that obscene war, they, with Job-like resolve, again found love and built new lives, new families. They do not deny the pain of their losses, but within the acknowledgment of those memories, they have created a new purpose in the part of their world that remains. The faults of life remain, but so do the virtues. And that is the essence of all lasting love. Nothing and no one in this world is perfect. Seeking perfection is to seek the transience of infatuation. The key to lasting love is to pay strong and continuing attention to the virtues even as you acknowledge the faults. The potential to love is the most profound bequest bestowed by the Creator upon humanity.

Would Job have been furious had he learned the origins of his grief—the Divine wager with Satan? Quite possibly not. By the conclusion of the book, Job has acknowledged that no human can comprehend the majesty of the universe or fathom the intricate workings of nature or the beauty that existence bestows upon its inhabitants. Awareness of the wonders of creation, even though he had experienced its flaws, lessened Job's frustration at his unearned tragedy, opened a path to rebuild a shattered life, and enabled him to again seek purpose in living.

Having the God-given right to argue with God, to demand righteousness in the world, does not restore the physical losses of misfortune, but having that right to argue with God empowers us to bring our complaint to the source of all creation with awareness that it will be heard. And though this does not necessarily ameliorate our physical pain, it does weaken the mental bind that tragedy can bring, allowing us, as did Job, to move beyond the past. We've learned from Job that in God's world pain does not necessarily imply guilt. The joy and the grief we encounter on our journey are not always the direct workings of God, which of course means that we too are partners in the making of our world.

EIGHT

Life and Death

Two Perspectives on One Reality

We learned in the previous chapter that God offered Abraham what seems to be only a marginally attractive covenant. "If you accept this pact," God tells Abraham, "I'll be your God. You will end your life in peace and tranquility. Unfortunately your progeny will suffer four hundred years of exile. But then the sun will shine, and their progeny will find liberation and reward." If this material world is all there is, then the promised reward would have to be vast to justify the centuries of travail. But if this material world veils a deeper and lasting reality, then getting God in the bargain would justify the promised travail. Abraham accepted the covenant, because God had taken him behind the veil in a vision in the night (Gen. 15:12–21).

Abraham was given a hint of the Eternal, a vision that life extends beyond the material world we see, that what we experience as death merely marks the end of our body's participation in life and not of life itself. If God as described in the Bible created a world in which our physical and biological existence is only a single chapter in an ongoing spiritual saga, then how we relate to life and death, and to God, must allow for our continued existence, though not one draped by a body. The God who created and then partnered with us in this world, our temporary residence, would then once again be our partner in that future world that outlives the physical.

But is such a transition, the physical giving way to the metaphysical, possible? I suggest that we not dismiss the prospect till we review the data. Even in the purely materialist world we know there are many exchanges, though not as radical, in which a totally unexpected reality emerges. If we had not seen it happen, who could have believed, watching a caterpillar build its enveloping cocoon, that from that same cocoon would emerge a butterfly?

The physical world invades and underlies all our senses. The metaphysical lies beyond our perceptual capabilities. And yet accepting the possibility of an ongoing existence need not be merely a leap of faith. Science has given us some quite firm clues that existence is very different from how we perceive it to be. At its most basic level even this physical world may be more ethereal than real. The discoveries of twentieth-century science have changed our understanding of reality. We've learned that the creation of the universe, the big bang, produced pure, exquisitely evocative energy concentrated within a miniscule speck, but not one iota of solid matter. And then that ethereal, massless energy condensed to become the solid material world we perceive. With the passage of time, it became alive, learned to feel joy and love, to compose symphonies, to reach for the stars.

Less than a century ago, as the understanding of nuclear physics progressed, we discovered that the solid matter we touch is really not so solid after all. In fact, it is almost entirely empty space. Scale an atom up to a size we can see. Let the central portion of the atom, the nucleus, be about the size of an orange. At this scale, the surrounding electron cloud will be four miles away. That's a ratio of one part solid nucleus to a million billion parts totally empty space between the nucleus and the electron cloud. This space is not an opening filled with air. Air is also composed of matter, but in a gaseous form. Those million billion parts are absolute void. Forces carried by virtual photons, virtual in that the effect of these photons is felt but the photons are never seen, hold the electrons in orbit and make solids feel very solid.

And then physics turns the tables on us, and we discover that even the diminutive particles that make up the atoms that make up our bodies and every other bit of the seemingly solid material in the world, the protons and neutrons and electrons, may actually be wavelets of information and not particles at all. The world may be the expression of totally ethereal information.

We sense the physical world because we are part of it. In the wave/particle duality that quantum mechanics has described for us and of which we and the entire physical world are a part, when we use an instrument sensitive to waves to investigate an item, we observe waves. When we use an instrument sensitive to particles, it is particles that we find in the identical item. Never do we observe both the wave aspect and the particle aspect simultaneously. The senses with which our bodies have been endowed—touch, smell, taste, hearing, and vision (even the photoelectric effects of vision)—are all particle detectors. In the making of our bodies, we have been set irrevocably in the particle mode, and that is how we view the world. The science of our time has revealed that other half of our reality. What would the world look like if, instead of acting as particulate entities, we interacted as extended, boundaryless waves? Perhaps the world would appear more like a thought than a thing, like an ethereal mind rather than a physical brain.

There is a real world, and we live in it. But as we study ever more closely the subatomic particles that join together to form the stones and flowers that we handle so casually, the world likens more to the shadows of Plato than the particles of Democritus. Our human perception of reality is a vast illusion, an artifact brought about by our limited senses of touch and sight. The scientific revelations of the past century have forced us to reevaluate the essence of that reality.

These data from the laboratories of quantum physics coupled with research in the functioning of the brain strongly imply that, sequestered within the material world that our bodies perceive, there is a metaphysical countenance not shackled to the physical. In each of us as individuals, our brain recognizes the transcendent facet of our being and refers to it as our mind. The quest in this chapter is to discover what happens to that brain-mind union once the brain no longer functions.

The idea that there is a nonphysical aspect to the brain-mind interface accords with our understanding of the act of creation itself. The big-bang creation of our universe was exactly that. Some totally metaphysical nonthing, either God according to the Bible or a quantum fluctuation according to NASA, created the physical universe from absolute nothing. The metaphysical became manifest as the physical. This being the case for our cosmic origins, positing a nonphysical (that is, a metaphysical) aspect for the consciousness of our minds is merely an extension of the scientific understanding of the totality of existence.

Of course, the thought that our consciousness resides within an ethereal entity separate from the physical brain has been the biblical claim for the past three millennia. "And when Jacob completed commanding his children, he gathered his feet to the bed and expired, and was gathered to his people. . . . And forty days were fulfilled, for such are the fulfilled days of embalming. . . .

And the Egyptians wept for him seventy days. . . . And his sons carried him to the land of Canaan [from Egypt, where he had died] and buried him in the cave of the field of Makpelah [in Hebron], which Abraham had bought . . . for a [family] burying place" (Gen. 49:33; 50:3, 13). In that cave Abraham and Sarah, Isaac and Rebekah, and Leah, (Jacob's first wife) were buried (49:31).

A careful reading of these verses finds an internal contradiction. Jacob died, was gathered to his people, and only months later was buried in the family crypt. If being "gathered to his people" implies his body or bones lying together in death with those of his deceased ancestors, then surely gathering and burial should have been concurrent. Yet here the gathering was separated from the burial by distance in the journey from Goshen in Egypt to Hebron in Canaan and by the months of time recorded in the Bible between death and burial. Clearly the "gathering to his people" relates to something other than the placement of the corpse in the grave.

Abraham, Isaac, Moses, and Aaron are among the biblical persons who were also "gathered to their people" long before their actual burials. In fact, every death and burial described in detail in the Bible follow this instructive yet seemingly illogical sequence.

In 2001, the prestigious peer-reviewed British medical journal *The Lancet* reported on a study of 344 cases of "cardiac patients who were successfully resuscitated after cardiac arrest." Of that number "62 patients reported near-death experiences (NDEs), of whom 41 described a core experience." The phenomenon of an NDE results when a patient, resuscitated after being clinically dead, describes in detail the acts that the doctors and nurses performed to save his or her life. In many of the cases, the patient experienced being greeted by deceased relatives and at times, though rarely, by deceased friends. Is this the biblical "being gathered to one's people," even as the body lies lifeless at its place of death? In the NDE, a bodyless consciousness views the "deceased" body as a separate entity.

Based on analysis of the medical data and interviews of the patients, the authors concluded, and reported in this highly respected journal, that the NDE is not the product of the radical chemical changes taking place in the brain at death, nor is it the result of psychological preconditioning. As such, the authors conclude that the "never proven concept that consciousness and memories are localized in the brain should be discussed."[1] The authors are writing in language that is scientifically acceptable. In plain terms, the "discussion" that they request relates to whether near-death experiences result from the persistence of a metaphysical consciousness that we associate with the mind even after the brain has ceased to exhibit the measurable activity we associate with life.

If the mind is not within the brain, not built of the body's flesh and blood, then it is not confined by the vitality of the brain, even though our body's awareness of the mind is totally dependent upon our brain as the receiver. At death, mind as a free consciousness would break from its relation with the physical body and brain. This release would account for part, or perhaps the entirety, of what people experience in NDEs. A parallel might be made with a radio and radio waves. Smash the radio and the radio waves remain unaffected. To hear the station's message, we need the radio as a receiver. For the body to get the message of the mind, we need the brain as the contacting receiver.

Werner Heisenberg, Nobel laureate in physics and one of the parents of all modern quantum mechanics, in his book *Physics and Beyond* suggests the philosophical implications of the metaphysical giving rise to the physical:

> Inherent difficulties of the materialist theory [of existence] have appeared very clearly in the development of physics during the 20th century. This difficulty relates to the question whether the smallest units of matter such as atoms [of which we and all objects are composed] are ordinary physical objects, whether they exist in the same way as stones or flowers. Here quantum theory has created a complete change in the situation. . . . The smallest units of matter are, in fact, not physical objects in the ordinary sense of the word; they are—in Plato's sense—Ideas.[2]

These are not the speculative words of science fiction. And don't let that exotic technical name "quantum mechanics" deceive you. Quantum mechanics is not some idle theory waiting in the wings for an esoteric application. Every time you turn on your TV, start your car engine, or place a phone call, you are putting into operation the insights of Heisenberg. The products of his mind are part of your everyday life.

The startling, totally counterintuitive, yet scientifically proven discoveries of physics reveal that our world, at its deepest level, is not built of tangible discrete objects. Rather, when we look closely, we find that reality is as gossamer as a thought, that existence is closer to being an association of ideas than a conglomeration of atoms. The dogmatic myth of materialism has been proven to be wanting, more fantasy than fact. Again, in the words of Nobel laureate and biologist George Wald, "The stuff of which physical reality is composed is mind-stuff. It is mind that has composed a physical universe."

The brain, built of flesh and blood, receives the mind's thoughts, just as a radio or TV receives the radiant broadcast signals and transcribes them into audible sound and perceived vision. The brain takes in information and stores and analyzes data. All our bodily senses are experienced in the brain, stored on the multiple maps of our bodies sequestered within our brains. We wiggle our toes and feel that motion in our brains, even though every bit of logic tells us we are feeling the motion in our feet. If you doubt this, just ask people who have had the misfortune of losing a limb and yet continue to feel the lost body part as the well-known "phantom limb."

The puzzle of the mind-brain interface is not in the recording and biochemical storage of the incoming sensory data. That is brain work. Specific regions of the brain are well known to be devoted to the processing of speech and vision. The paths of the incoming data have been largely identified. The puzzle is in the replay. There is no hint in the brain of how you hear or see what you have heard or seen. There is no sound in your brain. Put a stethoscope anywhere in the brain and all that is heard is the gurgling of the blood as it moves through the vessels. No voices. No music. But I hear voices and music. But where is an unknown.

The identical biochemical reactions that in one part of the brain store inputs related to the sounds we hear, in another location of the brain record the sights we see. But it is all chemistry and, even more perplexing, it's all the same chemistry. And yet from this chemistry emerge the immeasurably different sensations of sound and sight. But where are they? The pat answer is that we perceive these chemical reactions as sound and sight. Obviously that is how we perceive the chemistry. The location of that perception is the puzzle.

There is not a peep or whisper or glimmer of light anywhere in your brain. And even as you read these words, you're hearing them. The brain receives information from the body. From that information emerges the sensations of our emotions and the awareness of being our selves. The mind appears as a virtual reality, existing fully in another dimension, but manifesting itself as an emergent property of the brain. The brain calculates what the next move on the chessboard should be. The mind experiences the tension in making that decision. I am not suggesting here any radical departure from our current understanding of the world. Emergent virtual realities are fully a part of the concepts of modern physics. The most common of these would be the positive and negative charges that enigmatically emerge from protons and electrons.

We have no conscious awareness of the molecular processes that yield our memories. At times our thoughts seem to arise from nowhere. The wellsprings of consciousness—of how the mental sound of a long-gone melody emerges, of what forms the envisioned smile of a deceased parent—are a complete enigma. They are such a riddle that Robert Sapolsky, professor of biological science and neurology at Stanford University, was moved to write: "Despite zillions of us [neurologists] slaving away at the subject, we still don't know squat about how the brain works."[3] A similar evaluation was given a few years earlier by the avidly secular former editor of the journal *Nature*, Sir John Maddox: "Nobody understands how decisions are made or how imagination is set free. What consciousness consists of, or how it should be defined, is equally puzzling. Despite the marvelous successes of neuroscience in the past century, we seem as far from understanding cognitive processes as we were a century ago."[4]

The significance of these statements, appearing in the world's most widely read scientific journals, should not be underestimated. *Scientific American*, with its materialist view of reality, has no intention of implying a metaphysical underpinning for consciousness, a separation of brain and mind. This is the same journal whose editor-in-chief, John Rennie, in June 2002 penned an eight-page piece entitled "Fifteen Answers to Creationist Nonsense." This piece by Rennie demonstrates the materialist perspective of the journal and its editorial bias. With such a view of life, every effort is made to place the mind squarely within the flesh and blood of the brain.

Are all our musings, the sounds and pictures we find in our head, built of molecules, even though those molecules are silent and sightless? Or could there be a transition from a physical brain to a metaphysical mind? If mind is within the brain, it is very well hidden.

These concepts arising from investigations of the material world should force us to reassess the essence of our being. The ultimate reality of life and therefore the significance of death become very different from how our physical bodies experience them. Death signals a modification, really an alteration, in how we perceive life. In the words of Nobel laureate and physicist Erwin Schroedinger: "So in brief, we do not belong to this material world that science constructs for us. We, [our personal awareness of being ourselves], are not part of it. We are outside. We are only spectators. The reason why we believe that we are in it, that we belong to the picture, is that our bodies are in the picture. . . . And this is our only way of communicating with them."[5]

The paradigm of an unembodied consciousness becomes sister to the paradigm of an ethereal world sensed as real. The realization that mind gave rise to the universe, to use Wald's insight, finds strong support within the Bible. "How great are Your works, God, with wisdom You made them all" (Ps. 104:24). And then, "I am wisdom. . . . The Eternal God acquired me [wisdom] as the beginning of His way, the first of His works of old" (Prov. 8:12, 22). "Wisdom has built her house, she has hewn out her seven pillars" (Prov. 9:1). "With wisdom God created the heavens and the earth" (Gen. 1:1).

The Bible informs us, and Wald, Heisenberg, Schroedinger, Jeans, and Wheeler have come to confirm, that wisdom (skeptics may wish to substitute here the scientifically acceptable term "information") is the substrate, the basis, of existence. Wisdom is as fundamental to our universe as are time and space.

The old conundrum of how the physical brain gives rise to our ethereal and elusive consciousness, the sentience of the mind, has evaporated. To quote a portion of Sir James Jeans's words already given in chapter 4, "We are beginning to suspect that we ought rather to hail mind as the creator and governor of the realm of matter." Mind has been present in matter since the inception of the material world. Mind is ubiquitous in our universe, from the simplicity of the smallest atom to the complexity of the human brain. And because of this ubiquity of mind, the seemingly inert earth—rocks and water—could actually rebel against the demand of the Creator (Gen. 1:11–12).

The sentient mind of the human finds its source in the wisdom from which it was created, something along the lines of "The world thinks, therefore it is." Or, in a more localized description, as René Descartes wrote (ca. 1750), "*Cogito ergo sum,*" "I think, therefore I am."

Wisdom is parent, and matter the offspring. Hidden though it may be, this wisdom, the first of the Divine creations, surges forth to become the complexity and genius found in even the most primitive forms of life. It only awaits our discovery.

We are truly made (Gen. 1:26) and created (1:27) in the *tsel'em* of God. *Tsel'em,* the Hebrew word usually translated "image," contains the Hebrew root *tsel,* meaning "shadow." A shadow projects the shape of what casts it, yet it has no physical authenticity of its own. That notwithstanding, the presence of a shadow is readily perceived and its effect clearly felt, as we all know, seeing that we seek the comfort of a shadow on hot and sunny days. As Heisenberg, Schroedinger, Jeans, and Wald realized, we are truly the *tsel-em,* the ethereal projection, of the thought that brought existence into being. Each of us is a spirit clothed within a body.

Approximately nineteen hundred years ago, the biblical commentator Onkelos interpreted the subtle wording of Genesis 2:7 as conveying the definition of humanity. The verse relates that God put the soul of humanity (the *neshama* in Hebrew) into "the Adam, and the Adam became to a living being." The superfluous word "to," in "to a living being" (superfluous and therefore always omitted in the English translations; yet in the Hebrew it is present to teach) hints at a change from a lower to a higher form of existence. Based on this, Onkelos read the verse as "and the Adam became *a speaking spirit.*" Not a speaking animal or a speaking human, but a speaking spirit clothed in a physical body. The spirit of our *neshama* elevates us from being an animal that also speaks to a being whose primary essence is spiritual and only secondarily is physical.

When we consider the broad view of the evolution of life, how could it have been otherwise? The big-bang creation brought into being not matter, not protons, neutrons, and electrons, not elements such as carbon, nitrogen, or oxygen, not any of the other ninety-two elements we observe today. The big-bang creation brought into being energy fields. And from that created energy, the entire material world was constructed. The totality of the physical world, our bodies included, is made of the light of the creation.

If the discoveries in physics over the past century are correct, then that physically condensed energy of the big-bang creation is totally the expression of metaphysical wisdom (cited in Gen. 1:1) or information (J. A. Wheeler) or idea (W. Heisenberg) or mind (G. Wald). Physics not only has begun to sound like theology. It is theology.

Do you remember Casper the ghost? We never really saw Casper. All we saw was his white sheet moving about, causing a ruckus. Not surprising. Ghosts are invisible. Our bodies are the sheet. We, the true essence of our beings, are as invisible as a Casper—and as real as our thoughts. The quality of our lives here, how we relate to this world and most especially the concern we express for our fellow humans who are all also made in the image of God, vastly affects the quality of the part of our being that persists after the death of our bodies. Life becomes an episode in a saga of our making, a chronicle that reaches into our pasts and extends to our yet unformed futures. Life's experiences, joyful or tragic, form but a part of this far more grand continuum. Abraham realized this when he accepted God's offer of generations of exile to be followed by the reward of redemption (Gen. 15:13–16).

Both the rationalist theologian Maimonides and the esoteric kabalist Nahmanides tell us that God's relationship to an individual is based almost totally on how that individual relates to God. The more a person allows God to be felt in life, the more God will indeed be in that person's life. In the NDE of the medical world and in the Bible's telling us of being gathered to our people, God has given us strong indication that our lives continue after our bodies die. This enduring nature of our consciousness, whether it is housed within mind or soul, means that the Divine relationship that we build here on earth continues far beyond our physical stay in this world. As such, how we live affects far more than merely what may be written on our tombstones or spoken at our eulogies.

The words of the Bible and the discoveries of science as it probes the secrets of the white fire upon which the black fire is inscribed come to confirm that there is in fact a life after life. But what does God gain by informing us of this ongoing eternity? What can be the purpose of our having the knowledge that life doesn't end with the death of our bodies? In part it makes us a bit like Abraham, realizing that we live within and are a part of the web of existence that extends over all generations and all time from the creation. Abraham received this wisdom as a gift in a vision. We struggle to learn it through study and research. With this knowledge, God encourages us to relate to our joys and sorrows with an ongoing context in mind, just as Abraham did. As we will learn in the coming pages, God has made this physical chapter of life, within which we currently dwell, for our physical as well as spiritual pleasure.

NINE

The Desert Tabernacle

A Model for a Universe Built of Love

There's this tiny piece of real estate on our magnificent planet that's precious to much of all humanity. And for the past three thousand years, a large fraction of humankind has been fighting over who is to be its keeper. There's no water there, no precious metals, no energy-rich oil or coal. Just an abundance of spirituality. It is a place I try to visit each week.

Now topped by a mosque that was built over the ruins of a Crusader church, until just under two thousand years ago it was the site of the second biblical Temple. Just under three thousand years ago, it became home to the first biblical Temple, the one built by King Solomon. The Temple Mount today is as much a focal point for the jousting of world politics and aspirations as it has been in millennia past.

Psalms reveals that, beyond its being the esoteric embodiment of a universal spirituality, the physical plan of the Temple actually reflects how God interacts with the world: "Walk about Zion and go round about her, count her towers. Put your heart to her bulwarks; consider her palaces that you may tell it to later generations that this is God, our God forever. He will guide us till our death" (48:12–14). "This is God"? Clearly the Temple, as magnificent as it may have been, was not in any way God. The verse is better understood as teaching a very basic fact: the Temple and its functioning reflected how God functions on earth.

In a more subtle way, as we shall see shortly, the Torah, traditionally considered to have been written some four to five hundred years prior to the book of Psalms, teaches that the Tabernacle that accompanied the Israelites during their forty years in the desert also revealed characteristics of God's activity in this world. By examining the role of the Tabernacle, we continue to learn how God chooses to interact with Its creation.

Humans have a short memory when it comes to belief in the metaphysical. The people of the Exodus witnessed the ten plagues and the splitting of the Sea of Reeds (often translated "Red Sea"), yet just three days later, after these phenomenal demonstrations of God's power and control over nature, they were murmuring in rebellion (Exod. 15:22–24). By forty-five days after the Exodus, we hear, "Would that we had died by the hand of the God in Egypt, when we sat by the fleshpots, when we ate bread to the full; for you have brought us forth into this wilderness to kill the entire community with hunger" (16:3). And then a few days later, having had their hunger satisfied by a daily portion of manna from heaven (16:14–16), their thirst quenched by water from a rock (17:6), still the people questioned, "Is the Eternal God among us?" (17:7).

The constant flow of miracles notwithstanding, the people just did not get the fact that a caring God was active in their midst, even if not always obviously so. At this point God "realized" that a focal point was needed to do exactly that, to remind the people that God was actually always in their midst. In a sense it's surprising that God didn't appreciate this human need earlier. Or, perhaps more accurately, God was hoping that we humans would get our act in order and perceive the Divine reality. We didn't make that connection, and so God directed Moses to build the Tabernacle to fill the lacuna in our perception of the Divine. It was to provide a focus, a constantly visible reminder of God's presence and power.

Following the Exodus from Egypt, the Israelites marched to Sinai. Moses ascended the mountain and received God's instructions. Initially this consisted of the Ten Commandments (actually in Hebrew, the Ten Words; Exod. 20:1–17). Then follow, in the next five biblical chapters, the obligations describing how people are to act toward others and toward God (Exod. 21–25). At this point, as if confirming the need for a hub by which to anchor these duties, the Bible launches into a detailed description of the Tabernacle. So crucial are the particulars of its construction that the Torah repeats the plan, in part or in full, five times. In places it reads as if God were describing the making of a universe. In fact, God was doing exactly that, but in simile and metaphor.

As with any building project, the first step was to raise the funds for its construction. Moses was instructed to take up a collection, but not just from anyone. There was a powerful constraint on who could give: "And the Eternal God spoke to Moses saying, 'Speak to the children of Israel and take for Me an offering; from every man whose heart makes him willing take My offering' " (Exod. 25:1–2). The desire to give with a loving heart had to accompany the gifts. Twice the Torah adjures us not to appear at the Temple empty-handed if we hope to experience the Divine (34:20; Deut. 16:16–17).

Love was to be the first of the essential ingredients in this building dedicated to the Creator. The need for love pops up at every turn in the Tabernacle and Temple traditions. The Talmud informs us that God caused Jerusalem and the second Temple to be destroyed by the Roman armies because of lack of love, for senseless hatred, among the people. Prior to the destruction of the first Temple, the prophet Jeremiah cautioned: "One speaks in peace with his neighbor with his mouth, but in his heart he lies in wait for him. Shall I not punish them for these things?" (9:8). As prophesied by Jeremiah, God would facilitate the destruction because of the sly and deceitful relations rampant among the populace. God simply refused to tolerate the use of a structure dedicated to His name as a seat of duplicity and hatred even among biblical scholars who served at the Temple.

With the funds secured, God chose the project manager for the construction of the Tabernacle: "And the God said, 'See I have called by name Betsalel, son of Uri, son of Hur of the tribe of Judah'" (Exod. 31:2). No other character in the entire Torah has this designation of being "called by name." Clearly his name is itself significant.

A Talmudic tradition says that if, when naming a newborn child, the parents open themselves to the spiritual unity that pervades all existence, the name they select will reflect the potential of their child. Whether, as the child grows to adulthood, that potential is fulfilled is the choice of the individual, but the potential is there. The name Betsalel is a contraction of three Hebrew words, *Be,* "in" or "with"; *tsal,* "shadow"; and *el,* "God"; hence, "in the shadow of God." So well had his parents named him, and so truly had he lived his name, walking "in the shadow of God," that Betsalel was chosen by God to build the Tabernacle.

God then tells us that Betsalel is infused with "the spirit of God" (Exod. 31:3); the influence of the Divine is working directly through him. This trait, "the spirit of God," appears in the Bible very rarely and is so special that, in the entire Five Books of Moses, only three people are explicitly stated to have received it. Rare though it is, it makes its first appearance immediately following the creation of the universe, though not at that point related to a person: "And the earth was unformed and void, and darkness was on the face of the deep, and the spirit of God hovered over the face of the waters" (Gen. 1:2). "The spirit of God" is rare; "the spirit of God hovering" is unique, never again repeated in the entire Hebrew Bible. It was a onetime biblical happening. And the event that followed this singular occurrence was the emergence of light: "And God said let there be light. And there was light" (1:3). A onetime phenomenon, God's spirit hovered over the world, and from that came light.

Science doesn't like onetime forces. Incorporating them into a theory is too much like fudging the data to make observations fit a preconceived idea. Albert Einstein resorted to a fudge factor when, in 1917, using relativistic relationships, he derived a space-time description of the universe. Unfortunately, his equations indicated a universe in dynamic motion, a universe in either expansion or contraction. But this was patently in error. After all, in that day, we navigated according to the "fixed stars." The universe was clearly static other than local motions of planets and moons. To match the then current erroneous description of a static universe, Einstein introduced an arbitrary actor—that is, a fudge factor—the "cosmological constant," which balanced the dynamic terms in his equation and, voilà, a stationary world was the result. Barely a decade after this blunder, Edwin Hubble and Milton Humason, using the crucial Cepheid star relationships discovered by Henrietta Leavitt two decades earlier, presented solid data that the universe was indeed dynamic, in a state of rapid expansion. Einstein's fudge was removed, and the equation matched the data. Einstein could have predicted, on purely theoretical grounds, the most important discovery of the century, perhaps of the millennium. There was a beginning, a creation, of our universe.

The big-bang creation of the universe was a onetime event. Then immediately following its creation, the universe experienced its own onetime event, inflation, a sudden vast expansion of the universe from its initially compact state "not larger than the pupil of an eye."[1] Exactly what inflation was is still being debated, but some exotic effect set the universe onto its irreversible course of expansion. Estimates of our early cosmology indicate that within the first micro-micro-microsecond after creation the universe expanded, its space stretched by a factor of a million billion billion. From that initial burst of energy, the universe has been "coasting" ever since in its outward expansion. To give a scale for that number, that same factor of expansion would stretch a normal garden green pea to the size of the entire Milky Way, our galaxy, approximately 100,000 light-years in diameter.[2]

Following the inflation, the next main event in our cosmic history was the emergence of light. Torah tells us God's spirit hovered. Physics tells us inflation. And both tell us that what followed was light breaking free. The details are described by science as the emergence of light from the high-energy plasma within which it had been locked.

Is "the spirit of God hovering" actually an allusion to the theory of inflation? I would not affirm that absolutely. Yet the parallels are intriguing, and a tradition from Talmudic times established that the Tabernacle symbolizes the essence of our universe and its making.

God used the spirit of God to set the universe in motion. Betsalel received the spirit of God to make the Tabernacle. With this gift, the Talmud[3] tells us that Betsalel knew how to make universes from the letters of the Hebrew alphabet, the *aleph-bet*.[4]

Though other people in the Bible may have possessed the spirit of God, beside Betsalel only two biblical characters are explicitly identified as having been so endowed: Joseph as chief adviser to Pharaoh (Gen. 41) and Bilaam, the gentile prophet (Num. 22). Joseph, son of Jacob, who had been sold into slavery, refused the lascivious advances of his master's wife and was unjustly thrown into jail, where he languished for years, and still he remained faithful to the beliefs and values of his parents. Such strength of character might reflect the Divine spirit that Pharaoh, king of Egypt, detected in Joseph when, following Joseph's description of how Egypt might be saved from the impending years of famine, Pharaoh proclaimed, "Can we find one as this, a man in whom is the spirit of God?" (Gen. 41:38).

The Bible's revealing that God imbued Bilaam, a Gentile, with the Divine endowment of godly spirit and prophecy is very interesting. We learn here that though the Bible focuses on the successes and stumblings of Israel, the God of the Bible is the God of all humankind. Gentiles as well as Israelites have the potential for prophecy. At the closing of Deuteronomy, the fifth and last book of the Torah, we're told, "There never arose a prophet again in Israel as Moses, whom the Eternal knew face to face" (Deut. 34:10). Nahmanides asks and answers the question, Why this limitation? Why does the Bible add the stipulation "in Israel" to "no such prophet"? The answer is that in Israel there were no other such prophets, but in other nations there were also prophets, and Bilaam was the biblical example of a gentile prophet addressed directly by the Eternal God.

The prophet Amos lists in parallel three God-directed exoduses in one verse: "I brought Israel from the land of Egypt, and the Philistines from Caphtor, and Aram from Kir" (9:7). All nations have the potential for the prophetic experience. But to retain that Divine connection, they must actively tap the well of

Divine blessing. Though the Philistines and Aram each experienced an exodus at the word of God, we don't read much about them on the front pages of the daily press. They apparently chose to ignore the Divine mission with which they had been entrusted and so disappeared from the flow of history.

God, through these biblical episodes, reveals an interest that spans all nations. Divine favor is not limited to a single nation. As the prophet Isaiah conveyed God's message: "My house shall be a house of prayer for all peoples" (56:7).

The concept of the chosenness of Israel does not contradict this notion of the universality of biblical theology. To make the existence of a Divine plan conspicuous there needs to be a test group, a group that stands out. The people of Israel are it. Why Israel and not some other group? Amos implies that others were given the opportunity, but rejected it. Or perhaps what Rabbi David Hartman says, with such flair, holds the essence of the truth: "God fell in love with Abraham and Sarah and got stuck with their grandchildren!"

A simple analogy for the need of a "chosen" people might be as follows. Someone invents aspirin and gives it to a hundred people who have headaches. The next day there is not a headache among them. What has been proven? Nothing. The weather changed or their diet changed, and that is what cured the headaches. But if a hundred people have headaches and fifty take aspirin and fifty take placebos, and the next day the aspirin takers are cured, but the other fifty still suffer, then the effectiveness of the drug has been demonstrated. Israel is a test group, for better or for worse. And, indeed, the people of Israel do stand out, as they seem to defy the general flow of history. Which is exactly what such a group would be expected to do.

The ultimate fate of Bilaam brings a deep message concerning the level of free will that God has implanted in humans. Bilaam was a prophet with whom God spoke (Num. 22:9, 12, 20; 23:4, 16). God had told Bilaam that the people of Israel would be blessed and not cursed. But Bilaam wanted very much to curse them, and so he rebelled against God, actually went against God's command. The result of his defying God was that in a later incident he was "killed by sword in battle" (31:8). The measure to which God has allowed our free will to operate is extraordinary, even unnerving. God let Bilaam go the way he so desperately wanted.

When we want something very strongly, even though in our hearts we may know it is wrong, as did Bilaam, God lets us travel the path we have chosen. For a parentlike God that must be very painful, knowing in advance that the route chosen is destined to lead toward trouble. But that is the model, the game plan for the universe that the God of the Bible has chosen. In this world, God explicitly states, "Life and death I have placed before you. . . . Choose life that you may live, you and your offspring" (Deut. 30:19). Bilaam and the Philistines and Aram suffered from muddled judgment that caused them to make the wrong choice.

Betsalel didn't make that blunder, and so he became the builder of the Tabernacle. Immediately after we are informed that he was given the spirit of God, we are told that God imbued him with three types of cognition, *hak'mah, tevunah,* and *da'at* (Exod. 31:3). *Hak'mah* is the type of knowledge you get from reading a book or hearing a lecture. You understand what you heard or read and can recount what you learned. *Tevunah* is the ability to weave together information from different sources and arrive at something new. You want to build a bridge. For that you need to know the tensile strength of steel, the compressive strength of cement, the geology of the site, and how many people are expected to travel daily over the bridge, among other things. Weave this knowledge together, and you will be able to make a viable bridge. *Da'at* is the "Eureka" sensation. It's when you get an idea, but say to yourself, "I haven't a clue where I got that from." Some claim it's the Divine spirit that thrusts forth the idea. Louis Pasteur told us that chance favors the prepared mind.

As might be anticipated, the Bible reveals that God used these three types of cognition to build our universe: "The Eternal God with *hak'mah* founded the earth, with *tevunah* established the heavens; with his *da'at* the depths were formed" (Prov. 3:19–20). The fact that Betsalel was infused with the same four traits that God used to make the universe—the spirit of God and *hak'mah, tevunah,* and *da'at*—establishes the tradition of the Tabernacle as a model of the universe.

In five separate passages, the Torah lists the many items contained in the Tabernacle. The order in which the items are listed varies in each of the five, but the first two items are always the same. Consistently, the first item is always the Ark of the Covenant. Since the Ark contained the tablets upon which were inscribed the Ten Commandments, it represented revelation, the foundation of biblical religion and the Hebraic tradition. Without revelation, the Tabernacle would have no meaning.

Included in the list of items are a spice altar, whose pure scent rose, as do our prayers, heavenward; an altar upon which animal sacrifices were burned, a symbolic gift to God; a candelabra, in which oil was converted into an ethereal flame, a hint of the light of creation, which so long ago gave rise to the tangible material world; and a laver, or wash basin, for physical cleanliness. But the Torah chose none of these as the second item in the list. The second most important item is a bit of a surprise. It is a table upon which twelve loaves of unleavened bread were placed each Sabbath.

The table and the bread represented the physical abundance and the potential for physical pleasure provided by the universe.

By making the table and bread consistently second in importance after the Ark, representing Divine revelation, the Bible may be communicating a unique religious message: God wants our pleasure. Experiencing God requires satisfying and sanctifying the physical with the same intensity as sanctifying the spiritual. It is the road to what God via the Bible defines as holiness. Contrary to philosophies of self-renunciation, the Bible requires first of all, love, not denial, of one's own person. And that implies personal enjoyment of the material wealth provided within the creation by the Creator.

The tendency to see the world as a duality, to separate the material from the spiritual, is purely myopic. Though we habitually relate spirituality to prayer or meditation or some magnificent view of nature, the God of the Bible finds a broader ground for Divine inspiration. The very fact that the Bible five times over describes the construction of the Tabernacle teaches that the works of our hands have the potential to inspire.

The reality of life is that we have been imbued with a physical body within which, during our term on earth, resides a searching soul. Both body and soul require our personal efforts if they are to be adequately nourished: "Not by bread alone does a person live, but rather by all that comes from the mouth of God does a person live" (Deut. 8:3). "All that comes from the mouth of God" unequivocally binds the heartening spirituality of revelation with the physical satisfaction of the material world. An eternal spirit residing within our material bodies reflects a complementarity quite similar to that of which quantum physicist Neils Bohr spoke when describing wave/particle duality: these are two aspects of one reality that seem to be mutually exclusive, and yet both are necessary for a complete understanding of that reality.

For the soul to reach the highest levels of spirituality, the body must also do its part. A joyful body can take the soul to spiritual levels that the soul can never reach alone. Concisely put, abject self-denial is unmistakably not a part of biblical religion. Item number two in the Tabernacle, the bread-laden table, makes physical pleasure an integral part of the metaphysical journey.

We grow up embedded within the marvel of existence. By the time we are old enough to consider the wonder of nature, nature is already "old hat" to us. We've been experiencing it from the day we were born. Our constant familiarity with existence throws a cloak over the wonder and makes it appear mundane. Juxtaposing the metaphysical revelation as symbolized by the Ark with the material abundance symbolized by the bread-laden table exposes this deepest secret of our universe, the seamless junction of the metaphysical with the physical. So fundamental is this union that, again in the words of Nobel laureate de Duve, "one may legitimately wonder to what extent the success of life is actually written into the fabric of the universe."

The Tabernacle of the desert and, later, the imposing Temple on Mt. Zion in Jerusalem were far more than mere structures providing a place for worship. The essence of their construction and the means by which they were constructed tell us of a God that cares for all peoples and favors all peoples with the potential for prophetic wisdom, bestowing upon Its creatures a world steeped in spiritual and physical abundance. How unfortunate that the perception of the God of the Bible is often exactly the opposite: a God conceived of as tribal and mean-spiritedly stern. Kenneth C. Davis, in his book *America's Hidden History,* characterizes the early American puritan ethic as the constant fear that somewhere someone might actually be enjoying life.[5] The hard-working, even valiant, pilgrims were strongly committed to their concept of biblical religion. Sadly and unfortunately for them and for the development of Western culture, they missed the underlying pleasure message of the Bible. Almost four hundred years later, the misperception remains, as evidenced by the title of Christopher Hitchens's bestselling book *God Is Not Great*. The subtitle tells it all: *How Religion Poisons Everything*.[6]

God created this world with a wisdom that realizes the pleasure of the body can take the spirit of the soul to heights the soul can never reach on its own. This is the great message of the Tabernacle and an important quality of the dynamic God.

TEN

Knowing Truth in Your Heart

A Tale of Love

An ancient Talmudic tradition claims that we know the totality of all ultimate truths. It's a legend that was once taught to all children. It goes like this. God sends an angel to teach every unborn babe, while in the mother's womb, all the secrets of the universe, those of the heavens and those of the earth. Angels, being God's own messengers, transmit the information in an unerring manner. Then just before birth, the angel kisses each child just above the upper lip. That makes the slight mark each person has right below the nose. With this symbolic mark, God informs us of the universality of God's care for Its creatures. All humans are privileged to have this Divine wisdom.

The angel's kiss erases all conscious knowledge of the lessons, but leaves the information tucked within the subconscious. When we hear a profound remark and for some reason it has the ring of wisdom, our subliminal appreciation of the ultimate truth has surfaced. It's the feeling, "Haven't I heard that somewhere before?" We have, but it was in the memory of our soul, not our brain. Whether or not the tale of the angel's lessons is true, as the sage Rabbi Noah Weinberg teaches, the effect that such a legend has on a child's understanding of God's ways is tremendous. From childhood we learn that all people have this Divine knowledge of the creation.

There's a knock on the door, and a toddler hugs Mom's leg as she opens the door. There stands a specter of a human, wild and unkempt. Just as the child is about to let out a scream, the toddler sees the mark under the visitor's nose. "Mom, you mean he also knows God's ways?"

Now compare this tale of the angel's kiss with the message of a story that every child in the West hears, "Little Red Riding Hood." Don't even trust your grandma. First check out the size and shape of her teeth! The puritan ethic considered the body an obstacle to godliness, and fairy tales such as "Little Red Riding Hood" teach us not to get too close to other people. Biblical theology teaches that God wants the exact opposite.

The command to love your neighbor as yourself (Lev. 19:18; Mark 12:31; Matt. 22:39) has tiers of meaning. First, love is a biblical obligation. If it were not, then we'd be told "to try to love your neighbor." The God of the Bible claims we can actually learn to love. Too often we equate love with infatuation. We can't force ourselves to become infatuated with another person. But love can be learned, and not surprisingly, since God wants to engender this love among us, the Bible hints at how to succeed in that quest.

Biblical Jacob wanted to marry his first true love, Rachel. But his wily father-in-law tricked Jacob, and he married Rachel's sister, Leah, instead. After the traditional weeklong celebration of the marriage, Jacob was finally permitted to marry Rachel. And with that, we are informed, "and he (Jacob) also went unto Rachel and also loved Rachel" (Gen. 29:30). Why "also loved"? Jacob always loved Rachel, so why is the Bible telling us that he "also loved Rachel"? The "also loved" comes to teach that in the week Jacob had spent with Leah, he had realized Leah's virtues and in doing so learned to love her as well, even though Jacob was aware that Leah, like most of us, was not perfect: "Leah's eyes were weak" (29:17).

"Hear, Israel, the Eternal our God, the Eternal is One. And you shall love the Eternal your God with all your heart, with all your soul, and with all your might" (Deut 6:4–5). That love should be the central point of all biblical goals, second only to the understanding of the Unity of existence, is quite simply extraordinary. These two primary verses of the Bible tell us what the God of the Bible wants from and for Its creations. Love implies society, not isolation on a mountaintop abstractly contemplating the meaning of life. The biblical statement, "And God said, 'It is not good for man to be alone' " (Gen. 2:18), speaks of the centrality that human relationships occupy in God's plan of the world. Similarly, "Behold how good and how pleasant it is when brethren dwell also together" (Ps. 133:1). Together. We fulfill one of God's basic designs through our relations with others.

An exquisite Talmudic tale, which I first heard told by Rabbi Noah Weinberg, reveals the stunning Divine notion held within the concept of friendship. It elucidates why the verse commanding us to love, "You shall love your neighbor as yourself," closes with "I am the Eternal God."

It was a time, a few thousand years ago, of extreme political tension. The vast Greco-Roman empire was about to split into two mutually jealous factions, one part based in Rome, one in Damascus. Marcus and Aristos had grown up together in Rome.

In their youth they had been study partners in Bible class. They continued that tradition into their adult lives when fortune would find them in the same city, but that was rare. To provide for his family, Aristos had moved his family and business to Damascus. Marcus remained with his family in Rome. The two men traded the goods produced in their respective locations. They were a natural team. Their shared language and, more important, their shared trust were total. Twice yearly they would meet, once in Rome, once in Damascus.

It was early spring that a fateful voyage brought Marcus to the Mediterranean seaport of Beirut. From there it was a pleasant day's carriage ride to Damascus and the home of his friend. Their greetings were cut short by a pounding on the door. Syrian Greek guards burst in, seizing Marcus, brandishing papers accusing him of spying for Rome. The charge was ridiculous. "Tell it to the king," was the reply. The king was mightily unimpressed by the protestations of Marcus. Treason meant death by hanging.

Tried and convicted, Marcus asked only one favor. "Allow me to return home so that I may bid my family farewell and set my affairs in order. Then I will return to face the sentence. I am a businessman. I owe creditors, and others owe me."

The king smiled at Marcus's request. It was too ludicrous to be taken seriously. "If you return to Rome, we'll never see you again in Damascus, where the hangman's noose waits for you."

"I'll give a guarantee that I will come back. Aristos, my friend, will guarantee my return."

"And if you do not return, we'll hang him?!"

"Yes, he's agreed because he knows I'll return."

"Be back in twenty days, or he hangs at dawn on the twenty-first day."

Strong well-directed winds cut a day off the sea voyage from Beirut to Rome. There Marcus instructed his wife and children how to continue his business and hoped for their happier future. Amid heartbreak and tears, Marcus set off for the return trip to his death. To be certain of his timely arrival in Damascus, he allowed a week for what was usually a four-day journey by ship. The early summer breezes smoothly moved the ship out of the harbor into the Mediterranean. Two days into the trip the breezes weakened, and travel slowed. And then the worst occurred. The sea became calm.

A completely windless sea is a magnificent phenomenon, only truly grasped by those who've experienced it. Many years ago, while I was active in oceanographic research in the Caribbean, our ship was becalmed. The sea becomes a glaze, a plain of pure liquid glass. There is nothing to break the still, smooth surface, no waves, no ripples. At sunrise and sunset, just as the disk of the sun cuts the horizon, a deep ruby streak flashes, racing along the mirrorlike sea surface from the horizon to the hull of the boat. You bathe in the glow.

But for Marcus, the beauty was the symbol of a curse. The days dragged on. At last storm clouds appeared. Winds filled the once flaccid sails. It was as if the calm of the previous days had acted as a reservoir for the energy of the impending gale. The boat virtually flew through the waters, unfortunately to no avail. Beirut harbor came into view just as sunrise of the twenty-first day broke upon the city. In the fullest meaning of the word, the deadline had passed. The worst was about to happen.

Marcus leaped ashore, threw the skipper a pouch containing the final payment of his passage, hired a team of horses, and raced to Damascus. Swathes of white foam streaming from nostrils and mouths, the horses brought the carriage careening into the central square of Damascus just as the king's crier called out the news that a hanging would take place within the hour. The king, to publicize the foolishness of Aristos and his guarantee, had postponed the hanging till noontime, when he'd be assured of a large and eager crowd to witness the event.

Marcus threw himself from the carriage and ran headlong up the steps of the gallows platform. There Aristos stood, hands bound behind his back, a noose loosely draped around his neck. Marcus hugged his colleague, at the same time telling the hangman that he, Marcus, the traitor, the intended victim, the man convicted of treason, had returned. He must now stand in place of Aristos. That was obviously correct. The hangman began the exchange of the noose only to hear Aristos complain that it was not to be so. Marcus had guaranteed that he would return by the twentieth day and it was now the twenty-first day. Marcus had failed in his promise, and so he, Aristos, must hang. That made sense to the hangman until Marcus pointed out how ridiculous this was. He, Marcus, was the traitor, not Aristos, regardless of the guarantee. So he, Marcus, must hang. What logic is there to hang Aristos and in doing so let the traitor go free? That made sense to the hangman until Aristos countered that a guarantee was in effect and according to the agreement with the king, the original trial was no longer of significance. That made sense to the hangman until . . .

And so the argument cycled until the hangman, in bewilderment, retreated to the king's quarters to get his majesty's decision. The king, after listening to the hangman's tale, demanded time to ponder the case in silent meditation.

Moments passed. The fates of two men condemned, one for treason, one for friendship, hung on the will of the king. When the king finally came forth from his chambers, tears and a smile of anticipated joy graced his regal countenance.

"I've reached my decision," he said.

Aside from the brush of wind as it pushed its way past the gallows rope, there was only silence.

"I'll set you both free, but only on one condition. That condition is . . . you make me your third friend."

"You shall love your neighbor as yourself. *I am the Eternal God*" (Lev. 19:18). Through the Bible, God has informed us that when you can truly love another, not for what you can get but rather for what you can give, then a third partner joins with you in that friendship, the King of kings. According to the Bible, God tells us that if you want to build a loving relationship with God, start by loving other members of humanity, all of whom are made in God's image.

Friendship, how we relate to others, is the biblical measure of how we relate to God and how God relates to us. As bizarre as it seems, God wants our love and friendship. And that love and friendship can only be expressed through our relations with others. We are the spiritual image of God on earth.

God had commanded Abraham, the first of the biblical patriarchs, to circumcise all males of his household, himself included (Gen. 17:10–11). Abraham was recovering from the circumcision when God paid him a visit: "And the Eternal God appeared [to Abraham] by the trees of Mamre, as he sat in the tent door in the heat of the day." In the next sentence, however, "Abraham raised his eyes and saw and behold three men were standing near him. And when he saw them he ran to call them . . . and said, 'My sir, if now I have found favor in your eyes, please do not pass away from your servant. . . . Recline under the tree and I will bring some bread. . . .'" (18:1–5).

Abraham's concept of propriety is astonishing. Abraham is in the midst of a visit by the Eternal God, when he breaks off the Divine encounter and runs to greet three total strangers, offers them drink and food, and then serves them. Abraham ranked welcoming strangers to his home above communicating with God, and this received God's approval. A few verses later the

Eternal says, "Abraham shall surely become a great nation, and all the nations of the earth shall be blessed in him. For I have known him . . ." (18:18–19).

Our potential for love of God is elaborated in the Song of Songs. The poem's passionate affection between loved ones is also a parable for our love of God. The name of God appears nowhere in that poem, and yet a first-century scholar, Rabbi Akiva, declared that, although all the writings of the Bible are holy, the Song of Songs is the holiest of all.[1] Love of God remains a meaningless abstraction until it is anchored in a love for one who was created in the image of God.

The wording of several biblical passages reveals the extent to which even an agent of God will go to maintain affections within a family. Divine messengers have come to bring happy tidings to Sarah and her husband, Abraham. Sarah, from within the family's tent, overhears the angelic messengers telling Abraham that she shall bear her first child the coming year. That was quite a piece of news considering that she, Sarah, was eighty-nine and Abraham ninety-nine. "And Sarah laughed to herself saying, 'After I have become old shall I have such a pleasure? And also my man is old'" (Gen. 18:12). When God recounted Sarah's words to Abraham, God omitted Sarah's reference to Abraham's advanced age. "And God said to Abraham, 'Why did Sarah laugh saying, "Shall I truly bear a child, I who am old?"'" (18:13). We'd call it a white lie, a slight divergence from the full truth to maintain family peace. It's part of God's demand to be just *and* good. William Blake described this perfectly, but in the negative: "A truth that's told with bad intent, / Beats all the lies you can invent."

The message keeps repeating itself throughout the Bible. God wants our love, but wants it more than via the abstractions of prayer and meditations. Biblically, our love for God is most avidly played out in how we relate to others. The dynamic God of the

Bible, the God that told us, "I will be that which I will be," wants Its creatures also to be dynamic and proactive in forming a harmonious society. "You shall love your neighbor as yourself. I am the Eternal God." With God as the third partner, three is company, not a crowd.

ELEVEN

Understanding the Merciful God of the Bible

Golden Apples in a Silver Dish

Let's return to what King Solomon wrote three thousand years ago in the book of Proverbs: "A word well spoken is like apples of gold in a dish of silver" (25:11). But why should a well-spoken word, a word used in its proper place, be like a golden apple in a silver dish? The answer lies in the viewing. When observed from a distance, only the dish is visible. We see the craftsmanship and design, the beauty. But what the dish contains remains a mystery. Only when the dish is brought close are the contents discovered. Maimonides, in his book *The Guide of the Perplexed*, interpreted those words as they relate to understanding the Bible. The silver dish is the literal text of the Bible, as though seen superficially from a distance. The apples of gold are the deeper meanings held subtly within the nuances of the text and discovered only by a careful reading. Unfortunately these are often lost in translations.

Maimonides noted that Solomon wrote "apples of gold in a silver dish" and not "apples of silver in a golden dish." As gold is more valuable and beautiful than silver, so the hidden meanings of the words expand the import of the text far beyond that of a literal reading. A dish of silver is valuable. The golden apples in the silver dish are priceless.

Most people base much of their belief in and concept of God on the Bible's description of God. Yet a superficial understanding of the Bible all too often leads to naive and even erroneous notions of God's word and thus of God. The Bible is described as a song (or poem, the same word in Hebrew). "[And the Eternal God said to Moses,] 'Now write this song for them and teach it to the children of Israel'" (Deut. 31:19). As any student of literature knows, poems have many levels of meaning, some far deeper than simply the words themselves. God may be the subtlest of all poets, both in the wording of the Bible and in the structure and making of our marvelous universe.

Considering that the moral precepts prescribed in the Five Books of Moses once formed the basis for much of Western society, I think we'll find it quite instructive to explore some of the more maligned legal aspects of those five books. Just as the simplistic concept of a God who never changes course fails to match the Bible's description of God, so too the view of a harsh, unforgiving Deity as the God of the Bible fails to stand up to a close reading of the text.

Looking for depth of meaning in the Bible is no different than seeking the forces that are hidden in the most obvious facts of nature. I look to the east in the early morning and see the beauty of a sunrise. The day progresses, and so does the sun as it makes the westward journey toward sunset. A superficial reading of these observations is, "There goes the sun again, circling the earth on its daily voyage." Every human perception gives testimony to this "fact." Yet reality gleaned from a myriad of studies has revealed the very opposite. In the earth-sun system, it is the earth rotating on its axis every twenty-four hours that gives the perception of the sun's motion. That the earth's rotation is from west to east makes it appear that the sun rises in the east and sets in the west. But if this is true, there is cause to wonder. The circumference of the earth is approximately 24,000 miles at the equator. If you are standing at the equator, you make the entire 24,000-mile circuit each twenty-four hours. That means without taking a single step, you've traversed 24,000 miles in twenty-four hours. You're going 1,000 miles per hour, day in and day out, and there's not a single perception of this continual journey. The silver-dish reading of our world is that the sun is moving and the earth is still. The golden apple is just the opposite.

If you were told that in the beam of pure white light lies hidden the colors of red and green and blue, what a fantasy it would seem. White light is just that and nothing else. Then drops of water fall, and the white light passing through the droplets splashes into a rainbow of color. Isaac Newton is thought to be the one who discovered that white light is actually a mix of all the colors of the rainbow. Put those colors back together, and you get white light again. That is not so different from how we view the world. We exist downstream of the droplets. We see life and death, stars and galaxies, and time and space. We see the shattered light, the diversity brought about by the big-bang creation of nature. If we envision back through the droplets, we sense the oneness that underlies this world, the potential whole that each of the parts represents. Physics has taken us along the path by discovering that all matter is composed of a single source, the energy of the big-bang creation. The Bible goes a step farther and reveals that within this discovered physical oneness lies the spiritual unity binding all aspects of existence.

The world so familiar to our human senses is but a metaphor for a truth far grander than that of the most exotic complexity of life and brain. The diversity of the cosmos has arisen from a singularity not of the physical type, but of the spiritual kind, a spiritual unity brought into being as wisdom, the first act of the creation.

Notwithstanding the admonitions to temper justice with goodness (Deut. 6:18), to love your neighbor, whether Israelite or stranger, as yourself (Lev. 19:18, 34), to treat strangers, widows, and orphans fairly (Exod. 22:21–22), and to cleave to the Unity that binds all existence (Deut. 30:20), we're confronted with God allowing Cain to murder Abel (Gen. 4:8), with God's command that Abraham sacrifice his son Isaac (Gen. 22:2), with the much quoted "eye for an eye"—*lex talionis*—(Exod. 21:24), and in our time with the innocent born handicapped and stunted for life. At times it's hard to discern where the "good," whether of Divine origin or otherwise, makes itself felt within the "just."

Unfortunately, understanding the full implications of any one sentence of the Bible assumes that readers have knowledge of the entire Bible. Here is a clear example of the need to know the entire Bible before passing judgment on any one verse or thought therein.

Prior to entering the Promised Land of Canaan, Moses sent twelve spies, one from each tribe, to scout the land (Num. 13). Though God had promised to aid in the conquest (Exod. 23:27–28), the battle was to be fought by soldiers, not by angels. Strategic planning required on-site information. Ten of the spies returned with such a disheartening report of the power of the Canaanites that the men opted to return to Egypt. That error resulted in God's decreeing the forty-year trek in the desert. God realized that humans, and specifically the men, require a constant personal sign of the Divine presence. To provide this reminder, the men were told to sew "fringes on the corners of their garments . . . that when you look upon it you'll remember all the laws of the God" (Num. 15:39). Yet even the most devout Orthodox Jews have very few garments with fringes attached to the corners simply because a later elaboration tells us to "make fringes on the *four* corners of your coverings" (Deut. 22:12). Only four-cornered garments require the fringes. The law in this case uses the command in the book of Numbers to give the purpose of the fringes and the information in the book of Deuteronomy to describe its execution. Unless both passages are known, the understanding of this Divine directive is defective.

Probably the most misunderstood of all Torah precepts is the seemingly unforgiving demand of "an eye for an eye, a tooth for a tooth." The Bible gives this command twice. First: "If men strive together . . . and harm follows, then you will give life for life, eye for eye, tooth for tooth, hand for hand, foot for foot, burning for burning, wound for wound, mark for mark" (Exod. 21:22–25). Then a book later: "And if a man maim his neighbor, as he has done so it shall be done to him; break for break, eye for eye, tooth for tooth; as he has maimed a person so it shall be done to him" (Lev. 24:19–20). As we saw in the twofold description of the biblical requirement for fringes on garments, the fact that "an eye for an eye" is stated twice with slight variation implies a need for elaboration.

Immediately a logical inequality arises. A person with fine vision in both eyes nastily blinds the eye of another. Upon investigation, we discover to our dismay that the recently injured party had only one functioning eye prior to the injury. He is now totally blind. According to a simplistic reading of the law, the perpetrator of this evil act must now have one of his eyes destroyed. But can this be justice? The victim has been left blind while the aggressor still has fine vision in his one remaining eye. Hardly could this be called justice. Yet totally blinding the attacker, which would require destroying both of his eyes in retaliation, leads to the rather eccentric "an eye for an eye and sometimes two eyes for an eye."

In Exodus, a few verses after the admonition of eye for eye, we read the identical phrasing in regard to monetary compensation in the case of a dangerous ox goring another ox. "He [the owner] shall *pay* ox *for* ox" (21:36). Just prior to the repetition in Leviticus of the eye for eye law, we're told "and he who kills any person must be put to death. And he who kills an animal shall *pay*: life *for* life" (Lev. 24:17–18). One must *pay* life *for* life in the case of animals. We discover here that "life *for* life" means monetary compensation in cases other than murder. This is reinforced by what follows: ". . . breach for breach, eye for eye, tooth for tooth. . . . He who kills an animal shall pay, and he who murders a person shall be put to death" (Lev. 24:20–21).

That last phrase, "and he who murders a person shall be put to death," would seem to be superfluous. After all, if the law is literally eye for eye, tooth for tooth, then for murder, the penalty must be life for life. So why does the Bible need to specify the case of murder? The summary of these explanations, found later, clarifies the reasoning: "And you shall not take a ransom payment for the life of a murderer who is guilty of death. He shall be put to death" (Num. 35:31). Since we've already been informed of the supposed *lex talionis,* we do not need to be told that the murderer cannot buy off his punishment, that he must be put to death. And that is exactly the point. For murder no monetary compensation is adequate. All humans, whether citizen or stranger, rich or poor, are made in the image of God. No sum of money can replace the loss. But for all wounds, whether eye or tooth or limb, monetary payment commensurate with the injury compensates.

Besides the clarification of monetary payment for all injuries other than murder, what follows was truly a revolution in civil law: "One law there shall be for you as for the stranger as for the home-born, for I am the Eternal your God" (Lev. 24:22). One law for the stranger and the home-born; truly the night of dark oppression had given way to the dawn of hope. The God of the Bible in some matters may be unpredictable. After all, "I will be that which I will be" is God's choice of identification. You can know God only by God's acts. But with humans, there is a Divine consistency. All are human before the law.

And what of the remaining potential for capital punishment? Can execution ever be justified? First of all, accidental murder, known today as manslaughter, does not entail the execution of the one who killed in error (Num. 35:9–34; Deut. 4:41–43; 19:1–13). In the current State of Israel only one person has been executed, although many terrorists with lethal blood on their hands are in jail. Only Adolf Eichmann, who supervised the murder of millions of Jews, Gypsies, and physically and mentally challenged people, has suffered that fate. The biblical requirements for conviction of a crime are demanding.

We learn from the wording of the several biblical passages the restrictions that apply to witnesses of a crime. These relate to all crimes, including those that would result in the execution of the accused. The verses inform us that the accused must have been warned that his potential act was a transgression of biblical law and told the penalty for committing that particular crime and then observed in the act of transgression. And since the verses telling of this warning and witnessing are in the plural, we learn that there must be two or more witnesses (Deut. 17:6). These three components set the basis for all types of biblical convictions: warning prior to the act that the act is prohibited; knowledge of the penalty for the act; two or more witnesses to the fact that the guilty party had been knowledgeably warned; and two or more witnesses to the actual committing of the crime. The biblical restrictions for conviction are so severe that almost all trials would result in only a partial conviction. Yet the potential for capital punishment remains a biblical option. The Bible makes no apology. There are rules in this world, some given in nature and some given by Divine fiat. And there are consequences. It is the transgressor's bizarre choice. We all have the free will to thrust our hands into a blazing fire. Knowing the obvious consequences, we forego the option of choosing to do so.

The popular though erroneous idea of a brutal "God of the Old Testament" is further promulgated by the misreading of Jephthah's vow. For eighteen years Israel had toiled under the cruel oppression of the Ammonites. Jephthah, a powerful warrior, was enlisted to lead the fight for the Israelites against that very formidable oppressor. Realizing that he needed the help of God to defeat the enemy, Jephthah prayed, "If You will indeed give the children of Ammon into my hand, then it shall come that whatever comes out of the doors of my house to greet me upon my return in peace from the children of Ammon, it shall be to the Eternal God *veh* I will offer it as a burnt offering" (Judg. 11:30–31). The meaning of the vow rests on the vague Hebrew word used as a conjunction, *veh,* which can mean either "and" or "or." As is well known, it was Jephthah's beloved daughter who rushed out of the house to greet her father. Was the vow "and I will offer it as a burnt offering"? Or was the more likely meaning, "or I will offer it as a burnt offering"?

In this case the context should be clear. To begin with, the idea of offering a human being to God is abhorrent in biblical religion and absolutely forbidden. That was the basic lesson that Abraham had to learn as God stopped his murder of Isaac. Canaanites might sacrifice a human to Baal or Molech or some other local deity. Child sacrifice was a well-known and accepted practice of other cultures in those barbaric times: "For every abomination . . . they have done for their gods [Baal and Molech]; they even burn their sons and daughters in fire to their gods" (Deut. 12:31). It was these very cultures that the Bible came to replace. But such sacrificial acts were never directed by Israelites toward the Eternal God, and Jephthah had made his vow to the God of the Bible, not to Baal or Molech.

When Jacob fled from his home to avoid the wrath of his brother, Esau, Jacob vowed to God that if he would safely return to his parents' home from this exile, then "Of all that You shall give me I will certainly give a tenth to You" (Gen. 28:22). Well, Jacob had twelve sons and at least one daughter upon his return. Does it even enter anyone's mind that Jacob intended to sacrifice one of his kids? Obviously not; that would be ridiculous. So too with Jephthah and his vow. Clearly with Jephthah's vow, the *veh* means "or" and not "and": ". . . it shall be to the Eternal God *or* I will offer it as a burnt offering." Jephthah's daughter was consecrated not for death, but for a life dedicated to God: "And she knew no man" (Judg. 11:39). That too was unjust, but her father was the boss at the time, and she paid the price for his, as we shall see, bizarre understanding of the law of vows.

King Saul, in the heat of battle against the Philistines, vowed, "Cursed be the man that eats anything before evening" (1 Sam. 14:24). With the Philistines defeated, Saul learned that his son Jonathan had eaten some honey. Saul then told Jonathan that he "must surely die." However, "The people said, 'Shall Jonathan die, who has accomplished this great victory in Israel? God forbid. As the Eternal God lives not a hair of his head shall fall to the ground. It was with God's help that he performed today.' So the people rescued Jonathan, and he did not die" (14:45). Saul's vow clearly was made without a full understanding of its consequences. Hence it was annulled, as are all vows made without the full understanding of their consequences.

The error of Jephthah had nothing to do with sacrificing (murdering) a human, whether it was his daughter or anyone else. His error was in not rescinding his vow when he realized the full effect it had on a person other than himself. Biblically, God has decreed that a vow intended to harm another cannot be enforced. It is automatically cancelled, as if it never had been made. A person cannot make a vow that will unjustly affect another person. Vows relate only to the person making the vow. The God of the Bible does not give authority to people to inflict harm or inconvenience on others, whether by word or physical might. And if harm does occur, the Divine goal, we see, is to restore harmony and reconciliation, not incite a latent desire for vengeance.

Dante in his *Paradiso* gave the correct advice that Jephthah might well have followed:

To keep your vow, but be not perverse
As Jephthah once foolishly did to complete a rash decree.
Better a man should say I have done wrong
Than keeping an ill-formed vow, he should do even worse.

Perhaps the best instruction is not to vow in the first place. "When you vow a vow to the Eternal your God, you shall not delay to pay it. . . . But if you refrain to vow it will not be a sin for you" (Deut. 23:21–22). This reads as if vows are not at the top of list of what God wants from us humans.

We learn God's opinion of vows more clearly from the regulations related to the most extreme of vows, that of the Nazirites. Nazirites are those who take it upon themselves to deny, for a set period of time, the pleasures of any product derived from grapes, including the most pleasurable of those derivatives: wine and grape juice. God has created a bountiful world for us to enjoy. After all, the first humans were placed in the munificent Garden of Eden. Nazirites have relinquished a portion of that Divinely bestowed pleasure.

In the book of Psalms we read that "wine makes the heart of a person happy, making one's face as bright as oil; and bread sustains a person's heart" (104:15). God wants our relationship with the Divine to be joyful. The hair-shirt philosophy, that of self-flagellation and denial, is not the philosophy of the God of the Hebrew Bible. We are told explicitly to "Worship God with joy, come to Him with singing" (Ps. 100:2). So fundamental is this attitude about worship set in joy that in the Talmud there is an intense debate as to how to usher in the Sabbath, the most joyful day of the week. Should the blessing of the day be over bread or over wine? Bread sustains life. Wine makes life joyful. The decision was that, although bread is more central to survival, the joy of the Sabbath requires that the blessing of the day be made with wine. Yet Nazirites have denied themselves that pleasure. Because of having deliberately chosen to refuse this precious gift of God, when a Nazirite completes the term of denial, he or she must bring a sin offering for having transgressed a basic command of the Creator, namely, to enjoy the fruits of the creation (Num. 6:14).

In connection with the joyful blessing of the Sabbath day over wine, the Bible symbolically brings a Divine demonstration of the importance of human relationships relative to the seemingly unbending laws of the Bible. Since bread is more central to survival than is wine, when inaugurating the Sabbath over the wine, the bread should be covered so as not to "embarrass" or slight this most fundamental of foods. As we see in the following incident, to some not embarrassing bread was a very important concept—perhaps too important.

In an era when traveling posed serious risks from rogues and bandits along the way, at considerable physical risk to himself,

Rabbi Yisrael Salanter went from village to village in eastern Europe encouraging people to respect and love one another. His arrival in one of these small, isolated settlements on the eve of the Sabbath was an exciting event for the entire village. Families would vie for the privilege of hosting him. He would only stay at homes where the children had grown, so that there would be no possibility that he might be depriving the youngsters of the scarce foods available. In this particular village, the host family was selected at a meeting in the small village synagogue. Upon hearing the news, the husband ran to tell his wife of this honor. Rabbi Salanter would be eating and sleeping at their home, modest though it was. After evening prayers the townsfolk accompanied the husband and the rabbi to the door of the home. Wishes of Sabbath peace were exchanged, and the husband ushered the honored rabbi into his home.

The table was set; the wine and the bread were on the table; the wine goblet was polished. And then the husband noticed to his great consternation that his wife in her excitement had forgotten to cover the challah breads, the two braided loaves prepared especially for the Sabbath. The husband was aghast and let his wife know it in a very loud voice.

"Why didn't you cover the breads? You know you have to cover the breads before I make the blessing over the wine."

She cowered in shame and covered the breads with the embroidered cloth that they had used every Sabbath, all their many years of marriage.

"Why do you have to cover the bread?" Rabbi Salanter asked with an air of innocence.

At this the husband turned to his wife and again berated her. "You see, the rabbi thinks we don't even know this custom." Turning to the rabbi he said, "We cover the bread so as not to embarrass the bread as we bless the day over the wine."

At which this gentle rabbi replied, "You worry about embarrassing the bread after what you just did to your wife!? You are actually more concerned about shaming a loaf of bread than shaming your wife?"

Rabbi Salanter understood, and hoped the husband of the house would also come to realize, that although customs can be performed mechanically, perfunctorily, they often have within them the potential to demonstrate deeper truths. In this case, as it is so often, we are dealing with love. God wants us to love Him, certainly. But every indication in the Bible is that God can get along without our love if only we will exhibit love among ourselves.

Of the two basic names used in the Hebrew Bible for the Creator of the universe, as we have seen, *Elokiim* implies strict and unbending justice and *Ja/ko/vah* implies the compassion and mercy that Rabbi Salanter tried to instill in his host. Other clues to the nature of God come from other references to God's name. Directly after "I will be that which I will be," God tells Moses, "This is my name *le'olam*" (Exod. 3:15). The Hebrew word *le'olam* has three root meanings: "forever," "hidden," and "in the universe." "This is my name forever hidden in the universe." Considering the mischief that too often makes its ugly appearance, *le'olam* certainly rings true. At times God hides Its presence so completely that God becomes difficult to detect. God has that option, confusing and frustrating though it may be to us mortals.

The characteristics that we attribute to the Creator God of the Bible are very similar to the characteristics we attribute to the laws of nature. They are in effect at all times, in all locations, even if not obviously so. No physical object is immune from their influence, and their influence is the same for all objects. And although the laws of nature operate in the physical world, they are not physical themselves. This sounds very much like the biblical definition of God. But there is one key difference between the effects of nature and those of God. Nature has no attribute of mercy. Drop a cup on a stone floor and the cup shatters, even if that particular cup happens to be a family heirloom. In a totally natural world, when we perform an act, we get what that act brings, for better or for worse. It's totally logical, cause and effect. But the Bible claims that is not always the case. Repentance and forgiveness are also part of our world: "And it shall be when all these things shall come upon you . . . and you will rethink in your heart . . . and return to the Eternal God and listen to His voice . . . that then the Eternal God will have compassion upon you" (Deut. 30:1–3).

The power of the Eternal God brings into the world an aspect foreign to the laws of nature, the concept of mercy, the effectiveness of repentance in correcting for past acts and altering future consequences. Though the concept is totally "unnatural," the Bible assumes all peoples intrinsically know the trait of Divine mercy. We see this most clearly in the events recorded as the biblical book of Jonah.

God had commanded Jonah to warn the wicked inhabitants of the pagan city of Nineveh. Because of their rampant abominations, their city faced impending destruction by God. "And Jonah began to enter the city . . . and proclaimed: 'In another forty days Nineveh will be overthrown.' And the people of Nineveh believed God, and they proclaimed a fast and put on sackcloth. . . . [And the king of Nineveh proclaimed,] 'Let us cry mightily to God and let everyone turn from his evil way and from the violence in their hands.' . . . And God saw their acts that they turned from their evil ways, and God repented from the evil which He said He would do unto them and He did it not" (Jon. 3:4–10).

Notice that Jonah had said nothing about repentance, only that Divine judgment had been passed and that in forty days the Ninevites would reap the consequences of their past evil. The city would be overthrown. Yet the people of this pagan city knew that God is merciful, and so they fasted and prayed for mercy. But fasting and prayer were not sufficient. As the text states, "And God saw their acts *that they turned from their evil ways.*" In a sense the evil city was "overthrown," because that evil was replaced by moral responsibility. The people's repentance and corrective acts averted the severe decree. The God of the Bible may or may not be pleased by prayer and mediation. That is open for debate. But certain is the fact that this God, our God, is impressed by acts. Facts on the ground are what enter into Divine accounting. When God saw that they had changed their ways, "God repented from the evil which He said He would do unto them."

There's an interesting nuance in God's turnabout. Jonah initially rebelled against God's command to warn Nineveh. In Jonah's mind the inhabitants of that wayward city deserved to be punished. After all, they had been evil. Now they should get their "just rewards." Jonah did not want to warn them, lest they repent. Then the punishment would hit them like a bolt from the blue. To avoid fulfilling God's directive to warn the city's inhabitants, Jonah attempted to flee from God's presence. He booked passage on a ship bound for Tarshish, a port city toward the western end of the Mediterranean Sea. But escape from God is not so simple. Being God of the heavens and the earth, God brought a violent tempest, and Jonah was tossed into the sea. There he was swallowed by a huge fish and "held captive" in the fish's belly. Trapped and facing impending death, Jonah prayed to be forgiven. God responded and Jonah was saved. The fish spit him up onto the shore, and Jonah then headed off to Nineveh, finally complying with God's demand.[1]

Jonah's mixed responses present a conceptual problem. Originally he had fled from God's command that he warn the people of Nineveh of their impending doom. Jonah simply did not accept the idea that repentance should bring forgiveness of past sins. In Jonah's philosophy of life, when you do wrong, you get punished. Yet when Jonah disobeyed God and was subsequently cast into the sea, he was given the opportunity and time to repent while held captive in the fish's belly. Repent he did. "And he said, 'I called out of my affliction to God and He answered me; out of the belly of the depths I cried and You heard my voice. . . . The waters encompassed me even to my soul. . . . Yet You have brought up my life from the pit. . . . That which I have vowed I will pay. Salvation is of the Eternal God'" (2:3–10). Jonah's prayer is so eloquent, and the message of Divine mercy, mercy even for one who had argued against the very concept of mercy, is so significant to our understanding of God's way in this world, that Jonah is a main scriptural reading on Yom Kippur, the Day of Atonement, the only fast day in the Torah. The God of the Bible is simply not vindictive. And God wants us to embrace that humane trait. Hence in the same verse that demands, "You shall love your neighbor as yourself," we read, "You shall not take vengeance nor bear a grudge" (Lev. 19:18). God granted Jonah's release. Jonah got the mercy message and proceeded to Nineveh, carrying the Divine warning.

In the closing lines of the book of Jonah, Jonah, from the outskirts of Nineveh, watched the success of the people's repentance. God planted a gourd to shade him from the intense sunlight of the day, and then the gourd died. Jonah was furious. Then God put the situation into perspective: "And God said, 'You have pity on the gourd that you did not plant nor make it grow. . . . And should I not have pity on Nineveh, where there are more than sixty thousand people who cannot discern between their right hand and their left, and also much cattle'" (4:9–11).[2]

There is no brutality here, only commensurate compassion and forgiveness. When the people of Nineveh changed their ways, God changed the decree. Quite simply stated, the image of a vindictive, harsh, and vengeful Mosaic God, a God that a superficial reading of the Bible seems to demand, with no lenience or forgiveness, an eye for an eye and a tooth for a tooth, fails utterly with a careful reading of the text.

It is important to emphasize that Nineveh was a pagan city. Time and again, the Bible makes it clear that God's watchfulness and compassion extend to all humans. Biblical religion is not some elitist agenda, designed for the select set of members, allowing only those few a future life. Just like the advertisement of a few decades ago that proclaimed "You don't have to be Jewish to like Levy's Jewish rye bread," according to the Hebrew Bible, you don't have to be Jewish to be loved by God or to gain a place in the world to come.

Isaiah states this succinctly: "My house shall be called a house of prayer for all peoples" (56:7). That house is a house for all people, provided, as the book of Jonah makes very clear, the people do their part. The people of Nineveh did exactly that: "And *God saw their acts* that they turned from their evil ways." Nineveh is just one example of a way that God expects us to complete the task of perfecting the world that was set in motion at the close of those six evocative days of Genesis. "And the heavens and the earth were finished and all their hosts. And God completed on the seventh day His work that was done, and rested on the seventh day from all His work that was done. And God blessed the seventh day and made it holy, for on it He rested from all His work that God created to make" (Gen. 2:1–3). "God created to make."

God created this magnificent world and set it in motion, but much "making" remained and still remains to be done. Tragedy and injustice, the poor and the destitute are still to be found. Having been made in the image of God and endowed with intellect and free will, we can choose to partner with God in making from what was created a world that can at last fulfill the needs of all. That would be the fullest realization of God's compassionate plea for us to "open our hands to the needy" (Deut. 15:11).

TWELVE

Partners with God

Working with a God That Will Be

We've covered the nature of God according to what we've read in the Bible and also how we've experienced God in the world. The confluence between these two seemingly divergent sources of knowledge is there to see for all who wish to look. Yet the complaint we hear so frequently asks, If God is truly present and active in this world, why isn't that presence more obvious? Why is God so well hidden? The answer to the query is in essence the topic of this book's entire discussion. Let's glean from the preceding chapters a summary of what the Bible tells us about God's role in the creation It created. It can be a soul-enlightening journey, especially if we keep in mind that the same God that told us to choose life (Deut. 30:19) created a world in which earthquakes and tidal waves, plagues and disease snuff the breath of life from myriads of humans—and animals too—at times in the span of seconds. Is this the doing of the God about which the Bible teaches? Is God indifferent to our sufferings?

Before we judge to what extent the quality of indifference can be attributed to God, we need to focus on a simple truth. There either is or is not a God. On this there is no middle ground. We may debate what the putative God's role is in the creation, but on the existence of the Divine, it is either yes or no. The Bible obviously weighs in on the affirmative. And what is the evidence "on the ground"?

With the very opening word of the Bible we are informed that the "clay" of existence, the substrate from which every aspect of the world emanates, is a totally ethereal wisdom that finds its origins in the infinite wisdom of God. That something as intangible, as ephemeral, as wisdom could metamorphose and become the energy and eventually the solidity of the physical world is only slightly more incredible (and I use that word in its literal sense) than the scientifically accepted opinion that the energy of the big-bang creation ultimately became alive, sentient, and filled with all the emotions of love, joy, excitement, and boredom. Both positions, wisdom forming energy and energy materializing as life, appear to be true, once we realize that the wisdom in creation, as revealed by the meaning of the opening word of the Bible, *B'raisheet,* has been confirmed as the information (Wheeler), the thought (Jeans), the idea (Heisenberg), and the mind (Wald) discovered at the quantum level of every item and atom. We are truly the idea of the creation and, biblically speaking, the wisdom of the Creator.

Yet if Divine wisdom convincingly does form the basis of existence, we would logically expect a perfect world, with no errors, no unjustified tragedy. Perhaps we'd expect that, but that would depend on how God's role in this world is defined. If the world were a Spinoza-type creation in which God steps back after setting in place the finely tuned laws of nature, then there'd be no contradiction. Of course with such a world plan, there would be no reason for seeking a connection between God and humankind. And certainly that is not the biblical image of the creation. In the opening chapter of the Bible, nine times, with the statement "and God said . . . ," we have direct Divine direction in our cosmic genesis.

But then seven times we are informed that God considered this cosmic flow to be "good," even "very good." Shouldn't that goodness be obvious if God is in charge? Perhaps not. Being told so often that things are good sounds as if at times things were not so good. That would certainly be consistent with what we've learned concerning the fundamental attribute from which extends every act of creation. We've learned that to a limited degree the world has been endowed with the freedom to run itself, the *tzimtzum,* or contraction, of God's manifest presence inherent in the act of creation. After all, even the earth rebelled when it produced fruit trees in place of the Divinely mandated fruit trees that also yield fruit.

By Genesis 6, humanity has gotten so far off course that God states regret at having created humankind and for repair revamps the world in a way that reduces human life to a more manageable span of years. And even that does not fix the situation. God laments that the desire of a person's heart is "evil from youth." Did God just realize the problem? Couldn't God have gotten it right the first time? Or did nature rebel and produce a human not quite up to the required Divine specifications? That would be a replay of the earlier revolt of the earth and fruit trees. God said it was good. Conceivably that "good" might possibly have meant merely "good enough." Though our heart's desire is evil from our youth, God allowed us somewhat defective humans to proceed to populate the world. The potential for good probably exceeds that for bad, so God makes do and works with what He's got.

By human logic it is illogical that the reputed infinite God of the Bible would create a world with such basic flaws. Physicist

Steven Weinberg bemoans the fact that such a flawed picture of God is equivalent to no God at all. But we are dealing with the Bible's own description of God. So how can we argue that this is equivalent to no God at all? Fortunately Professor Weinberg, being an honest fellow, also informs us that, though he is willing to evaluate God's apparent absence, in this topic ("What About God"), he "leaves behind . . . any claim to special expertise."[1] I wonder why Weinberg entered the theological fray, since by his own words he is not qualified to do so. Would a person lacking detailed knowledge of theoretical physics dare to tell Weinberg that his field of research lacked credibility?

The source of potential calamity lies, as we have learned, in God's proclivity for withdrawing control, the *tzimtzum,* the contraction, of God's manifest presence. The surprise of twentieth-century physics is that this *tzimtzum* also applies to the laws of nature. Though God does not play dice with the universe, to paraphrase the statement attributed to Einstein, God does indeed allow the universe to play dice. By itself, as is couched within aspects of quantum uncertainty, at God's option, nature is allowed to run its own course. The Bible offers many examples, in addition to the few presented here, of how that leeway granted both to nature and to humans is played out in the world.

The Israelites are about to enter Canaan. To help, God promises to send in swarms of hornets to chase out the enemy "little by little." Why little by little? Why not all at once? The Bible informs us of the reason: "Lest the wild beasts of the field multiply" (Exod. 23:29; Deut. 7:20). God chooses to control the hornets, but not the beasts. God will help, but we have to finish the task. In general, for those situations where we can solve the problem by our own efforts, God relegates completion of the task to us.

The Israelites are about to enter battle. God informs the combatants that God will be with them in battle. Then immediately

God urges all who are engaged to wed but have not yet married, or built a new house but have not yet entered, or planted a vineyard but have not yet tasted the harvest to return home lest they die in battle before attaining their goal. Lest they die in battle? God just told these warriors that God would be with them battle. That sounds like a guarantee of success. So why the possible death in battle? The guarantee is for the army, the group, not the individual.

In God's management of the world, the Bible makes clear that the acts of an individual strongly affect the community. As we discussed, God led Joshua and the children of Israel through a rapid conquest of Jericho. But one of the soldiers, Akhan, stole silver and gold from the city's forbidden rubble. For that transgression, in the following battle for the city of Ay, Israel was soundly defeated (Josh. 7:1–21). One person sinned and many innocent people perished. The message is clear. We are our brothers' and sisters' keepers, individually and communally.

The world gets its share of free reign and when a mess arises, the God of the Bible may enter to aid in the repair. Nipping the potential evil before allowing it to flourish would be a compassionate world-management system, but that fails to match the blueprint brought by the Bible. The logic lies in the need for an unhampered free will. God hides the Divine presence sufficiently to allow each of us to make our own choices, for better or worse, freely within the confines of our physical and social landscape, without the specter of a cosmic Force peering over our shoulders, judging our every act. Dr. Joseph Hertz, former chief rabbi of England, describes the human situation perfectly: "Though man cannot always even half control his destiny, God has given the reins of man's conduct altogether into his hands."[2]

The word "Torah" in Hebrew has the root meaning not of "law," but of "teaching" or "instruction." The Torah accomplishes that task in two basic ways: by direct statements—do this, don't do that—and also by the descriptions of the lives and journeys of people addressed by God. As we've learned in this book, an examination of these narratives presents the impression that God is learning how to relate to the creation It just created.

God forms Adam alone and then informs us that it is not good that Adam is alone. This is the first acknowledgment by God that some aspect of the creation is not good. But then why have created Adam alone? When the two sons of Adam and Eve bring gifts to God, one gift is accepted, one rejected. That is a sure recipe for provoking jealousy in the rejected son. From the murder that followed the rejection, it appears that the potential for jealousy was unfortunately triggered. Initially, the people of the Bible lived to nine hundred years. This turned out not to be a good plan for people with the extent of free will allotted to them by God. God had to correct this by reducing longevity to about a century. God explicitly chose Saul to be the first king of Israel and then stated regret or reconsideration at having chosen him. God then had the kingship taken from Saul. Was God unaware of the impending consequences seemingly inherent in all these Divine decisions? Superficially that would be the reading of the text.

However, in the closing of the fifth book of the Bible, Deuteronomy, God in general terms tells Moses what the future will bring to the people of Israel: rebellion, exile, dispersion, and ultimately return. The implication is that God, being in one aspect outside of time, is cognizant of the broad-spectrum flow of the future. The specific nuances are in our hands. Rivers flow toward the sea, but some meander more than others. God did not have to reveal these diversions recorded in the Bible. Had they been deleted, we'd only see the outcome and would assume a world perfectly managed in every aspect. These "errors" are included to teach. Just how closely God micromanages the world is cause for much debate. Regardless of the answer, these Divinely induced changes in venue bring with them a vital lesson. Not only is God's management of the world at times relaxed. But if God, the source of the past, present, and future, can unabashedly respond to changing contingencies, altering plans to meet new realities, obviously so must we, who are made and created in God's image, be willing to do so, difficult though that may be.

The seeming vicissitudes of our Creator are found in the self-definition of God's name as told to Moses on Sinai, "I will be that which I will be." "I will be" is a future-tense verb indicating that God is active and ever defined by God's own acts in this world. Stasis in godly manifestations would be reassuring and predictable, but stasis does not meet the criteria of the biblical God. We can only know God by what God does. What God is is what God does in our temporally and physically limited span of existence.

Though the Bible, in a superficial reading, makes the case for male domination, a close reading of the text reveals that the input of both man and woman is important.

The English language does not differentiate between masculine and feminine for most words. Hebrew, like most ancient languages, retains the gender difference; all nouns are either masculine or feminine. The central role of the feminine reveals itself in the Divine inputs to the world—all are feminine in Hebrew. These are the Torah, the name for the text revealed or inspired at Sinai; the *nefesh,* which is the soul of animals, created on day five; the *neshama,* the soul of humans, created on day six; and the *shekinah,* the indwelling spirit of God. In Proverbs we are told that God created the world with three types of cognition, *hoamah, teunah,* and *daat,* knowledge, understanding, and insight—all three of which are feminine.

Though in Hebrew the default gender is masculine, when women spoke, their men listened. When the first of the patriarchs, Abraham, hesitated to follow the advice of his wife, Sarah, concerning Ishmael's unacceptable behavior, God told Abraham, "In all that Sarah says to you, listen to her voice" (Gen. 21:12). A generation later, Rebekah told Isaac that their son Jacob needed to find a wife more appropriate than one of the local women of Canaan. Immediately, "And Isaac called to Jacob and blessed him and charged him saying, 'Do not take a wife from the daughters of Canaan. Arise and go to Paddan-aram . . . and take a wife from there'" (28:1–2). And then again a generation later, Jacob, the third of the patriarchs, after serving his wily father-in-law as a shepherd for some twenty years, at God's suggestion Jacob planned to return to Canaan. But before finalizing that decision, "Jacob sent and called Rachel and Leah (his wives) to the field to his flock [of sheep]." He told them of the need to leave, and they responded, "Whatever God has said to you, do" (31:4, 16). A superficial reading of the Bible sees a very male-oriented stance in the biblical text. A careful reading reveals that at these three absolutely crucial junctures in flow of the Bible, junctures that shaped the destiny of the people of Israel, the woman's voice is clearly heard.

Once when the women's voice was not heard, the entire tribe suffered greatly. Just prior to entering the Promised Land of Canaan, twelve spies were sent to reconnoiter the land. They returned with a report that the land was indeed magnificent, but ten of the spies claimed that the inhabitants were giants and therefore unbeatable. The decision was made to return to Egypt, and the people cried, "Why did God bring us to this land to fall by the sword? Our wives and our little ones will be captives; it is better that we return to Egypt" (Num. 14:3). This statement marked the beginning of the forty-year trek in the desert. Note the reference to "wives and little ones." Clearly those opting to return to Egypt were the adult males, not the wives or the kids. Because of this, only the adult men died during the years in the desert. Other than Miriam, all the women survived. Had the women's voices been heard at that key juncture in Israelite history, the long travail in the desert would have been avoided.

We see that the relation of biblical characters to God, whether male or female, was not necessarily serene. The book of Psalms, which in a sense may be seen as love letters to God, reveals the almost schizophrenic relationship we have with our Creator. Psalm 22 says, "My God, my God, why have you forsaken me?" The very next psalm says, "The Lord is my shepherd, I shall not want." It would be hard to find two descriptions more divergent than these of the relationship between a lover and the loved one or between Lord and servant. These two verses in words capture the frustrations that biblical Job must have felt during and even after his unwarranted sufferings. Why did God "abandon" him? No Divine explanation was offered, only the assurance by God that God had never forsaken Job, even though God allowed Job to be subjected to horrific punishment. And as proof of God's shepherding, God ensured that Job survived those troubles. What Job learned and in essence taught the world is rather exasperating. With the way the God of the Bible relates to this world, not every ill wind that comes one's way is sent by God. The pain may be, from a Divine perspective, completely unwarranted. So why doesn't God step in? That is part of the Divine management system.

Biblically, there's evidence that for all the freedom of purpose implied in "I will be that which I will be," God has set ground rules, limits, not only for humans, but also for Divine behavior. Time and again, following the Israelites' exodus from Egypt, during their long trek in the desert toward the Promised Land of

Canaan, the people rebelled. In most cases God enacted a limited reprimand. But on two occasions, the insult to God was so severe that God voiced determination to destroy the entire tribe.

Immediately after their escape from Egypt and the pursuing Egyptian army, the Israelites marched directly to Mt. Sinai. Upon their arrival, Moses was summoned by God to ascend the mountain. According to the text, Moses stayed there for forty days, during which time he received Divine instructions, primarily the Ten Commandments. But forty days is a long time to be without any word from the leader who has taken full responsibility for this entire exodus. The people despaired of his ever returning and so felt the need for a new leader and a new symbol. That symbol was the infamous Golden Calf. Moses, being on the mountain, was totally unaware of the people's idolatrous transgression. God, however, saw the situation, at which point He vowed to destroy the entire people, telling Moses to leave Him alone so that He might carry out the destruction. Again, only months later, just prior to entering the Canaan, the people feared that they could not conquer the mighty inhabitants of the land, and so they opted to return to Egypt and slavery rather than die in battle. As at the incident of the Golden Calf, this act was so appalling to God that God once again threatened to destroy the entire Israelite nation.

In both instances, Moses "explained" to God that this would be counterproductive to God's ultimate goal of recognition among the nations of the world. "If you destroy this people, then the nations which have heard of your fame will say because the Lord was not able to bring them into the land which He swore to them, therefore He has killed them in the wilderness" (Exod. 32:12; Num. 14:15–16; Deut. 9:28). In each case, God acquiesced to Moses's logic. Later God admits to the validity of that human reasoning. "I thought I would destroy them; make their memory cease, but I feared lest the enemy say our hand is exalted, and not that God has done this" (Deut. 32:26–27). Abraham asked, "Shall not the Judge of the earth act justly" (Gen. 18:25). The Psalmist bargains, "What benefit is there in my blood, if I descend to the pit? Will the dust praise You, will it tell of Your truth?" (30:9).

There is a plan by which God interacts with this world. And one goal of that multifaceted plan reaches out to bring an awareness to all nations of God's concern for all inhabitants in the creation It brought into being. As the prophet Amos taught, not only did God bring Israel out of Egypt, but also the Philistines out of Caphtor and Aram out of Kir. Three "exoduses" are described in one biblical verse. Though that Divine connection may not always be obvious to, or in accord with, our limited human logic, the connection and care are there.

Most ancient cultures remove the troubled episodes of their histories from their records and preserve only the blessed portions. The Bible keeps it all, and in doing so shows a series of incidents that expose God's relationship to the world It created. Most important of all, we learn that God is present and interested in all nations and all peoples. The people of Israel may be a marker making more obvious God's active role in history, and that role is there for all to recognize. But that has not limited God's interest to this one people. The role that Israel plays is to be an indicator, an example, so that all people may recognize the Oneness that lies beyond the diversity of existence. We are all intertwined, as individuals and as members of the larger community, truly as our brothers' and sisters' keepers. Being created in the image of God, we are partners in the final making of the world.

We opened this book with the acknowledgment that the entire universe might not be Divine at all, that all of existence could conceivably be described as the product of an unthinking blip in the laws of nature, a random fluctuation in a timeless quantum field, ringlets of energy made manifest as they break free from an eternal reservoir filled with their potential. Psalm 19 foresaw the standoff between these two opinions: "The heavens declare the glory of God, and the vault of the firmament tells the work of His hands. Day to day gives forth speech and night to night expresses knowledge. There is no speech and no words; their voice is not heard" (Psalm 19:1–3). These opening verses tell us that the heavens proclaim with no equivocation God's glory. Then immediately we learn that nothing is heard. The message is there, but to perceive the presence of the Divine, we have to listen very carefully. God knocks very gently: "A great and strong wind fractured the mountains and shattered the rocks before the Eternal God, but God was not in the wind; and after the wind an earthquake but God was not in the earthquake; and after the earthquake a fire but God was not in the fire; and after the fire a still small voice. And it was when Elijah heard it that he wrapped his face in his cloak" (1 Kings 19:11–13).

When Steven Weinberg complains that "signs of a benevolent designer are pretty well hidden," he is right in line with the biblical description of God's role in the universe. There's no hint of a constant microengineering by God either in the world or in the Bible. As we learned, the very act of creation is God's *tzimtzum*, a lessening of Divine control, a withdrawal of evident presence. Rashi exposed and twentieth-century quantum physics confirmed that this lessening (Divine or natural) extends even to seemingly inert earth and atoms.

The God that most skeptics reject, a God with unceasing hands-on control, is simply not the God of the Bible. The biblical God may enter the fray when the flow of nature and humanity strays too far from the intended teleological path. In general, however, the running of the universe is not a power play by God. We and the laws of nature, which are themselves creations of the Creator, have a major role in the scenario. The Bible recognizes that flaws exist in nature's designs. It even describes them. The God of the Bible expects us to fix them. That's what partnership is all about. Not only are we our brother's and sisters' keepers, we are even God's keepers, as is God our Keeper.

Eventually the blemishes in nature and in society can be repaired. That is our part of the bargain. Famine and draught on a globe 70 percent covered by water only await a solution garnered from the genius with which we have been imbued, much as the devastation caused by earthquakes can be confined through prudent construction with steel-reinforced concrete.

And for humanly imbued tragedy, the biblical demand that one day a year we observe a total fast (Lev. 16:29) carries a powerful lesson. We can restrain our most basic drives—when we want to. The day of fasting moves the theory of our ability to rule our desires into an enacted fact. Biblical religion is littered with rituals, and most relate to life in the marketplace, not in the house of worship. Not by chance. Abstract theory is fine, but acts are what bring results. There's no difficulty being holy in a church, synagogue, or mosque. The challenge comes when we step outside and confront our fellow humans, some of whom do not conform to our standards. Remove all theological implications from the day of fasting and still it remains a meaningful exercise for atheists and theists alike.

The first biblical constraint placed on humankind related to that most primal human need, food—the forbidden fruit in Eden, an arbitrary limit on the desires of our free will. The first question asked of humankind by God was, "Where are you?" (Gen. 3:9). Among what fantasies are you hiding? What excuses have you concocted to justify the failure of humanity to fulfill its potential? Well, Adam blamed his wife, and his wife blamed the snake. There are a myriad of reasons for moral and ethical slovenliness. Just ask the contractors whose inferior work led, during the May 2008 earthquake, to the fatal collapse of the Xinjian Primary School in the town of Dujiangyan, China.

There is a horribly painful irony to that disaster. That small, ancient town of Dujiangyan in western Sichuan is well known to the Chinese. The first system built in Asia to control the cyclic disasters of flooding and drought, the Dujiang Dam, was built there almost two thousand years ago. The Dujiang Dam has been lauded by oriental scholars as an example of humankind's ability to tame the forces of nature. The potential for that ability is present, but when the pillars intended to support the upper floors of a school are filled with sand and not cement, the execution of that potential is absent. Guo Xiaoyan, whose only son was crushed in the school collapse, summed up the reality: "This was not a natural disaster. It was manmade."

In Steven Spielberg's classic *Schindler's List,* Oskar Schindler attempts to explain—not to justify—the barbarity of Amon Goeth, the German in charge of allocating slave labor to work camps. "In normal times" Schindler says, "Goeth would be like most people. He'd be okay. But war brings out the worst in people. Never the best. Always the worst." This statement by Oskar Schindler is totally amazing, considering that Schindler risked his life a dozen times over and expended his entire vast fortune to save the lives of the thousand-plus slaves who worked in his factory from the Nazi slaughterers. And thank God he succeeded.

Oskar Schindler was the living example of a very different truth. It is the strength of one's will that determines which way a person turns when faced with oppression or trials, whether they are induced by war or other circumstances. Weakness of character and the imperfect mores of a culture, not war, bring out the worst.

In Spielberg's epic there is an incident reminiscent of a situation in the book of Esther. Schindler urged the manager of another factory to save the slaves in his charge. The man refused, saying he could not take any further risks. In doing so he remained nameless, erased from the same pages of history upon which Schindler is inscribed.

In the biblical book of Esther, Mordecai asked Esther, his niece and now queen of the realm, to go before the king and plead to have the royal decree to destroy all the Jews rescinded. Esther wavered in her reply: "All the king's servants and the people of the king's provinces know that whoever, man or woman, shall come before the king within his inner court, who is not called, there is one law for him, namely, to be put to death; unless the king holds out his golden scepter; then he may live. I have not been called to come to the king these thirty days." In short, Esther was not about to risk her life to save others. Upon which Mordecai explained to her what might be called the realities of life: "If you remain silent at this time, relief and deliverance for the Jewish people will arise from a different source. But you and your father's house will perish" (4:11, 14). We read the book of Esther for one reason only. Esther realized the truth of Mordecai's charge and approached the king. There may have been a host of other queens before her who opted out. Their names, like the name of the factory owner who rejected Schindler's entreaties, are forgotten, blotted out from history.

Within the subatomic world, there is a probabilistic pattern established by the laws of nature. Individual quanta, however, may "choose" not to follow the given path. So too is the history of humanity. Tortuous though the trend may be, God has a plan for the world. The microengineering of that plan is largely up to us. There is a flow from pagan barbarity toward the elusive goal of peace on earth, goodwill to all. Each of us, as individuals, chooses whether to enhance or impede the flow toward the Divine goal.

Near the closing of the Five Books of Moses, we are told that "the secret things belong to God, but the things that are revealed are for us and for our children forever" (Deut. 29:29). "Secret things belong to God." As God told Job, there are limits to what we humans can know. Although we cannot discern all the reasons for every event, we can use our reason to partner with God in bettering the world. The Bible devotes one chapter to the creation of the universe, a mere thirty-one verses, and ten chapters to the construction of the Tabernacle that accompanied the Israelites during their forty years in the desert. The Bible is not so interested in how to get to heaven. In fact, there is no direct mention of life after life in the entire Torah. Our God-given goal is to make the world so perfect that we will have heaven here on earth. The prophet Micah brought the world the definition of true religion: "It has been told to you, humankind [*adam*, in Hebrew] what is good and what the Eternal God asks from you: that you perform justice, love merciful kindness, and walk in humble modesty with your God" (6:8).

Note the simplicity of the requirements for a godly life: justice tempered by kindness and humility. The only trait of Moses's character that is recorded in the entire Bible is that he was "the most humble of all men on the face of the earth" (Num. 12:3). Moses confronted Pharaoh, the most powerful ruler of the time. Humility is not the equivalent of being self-effacing. Humility is knowing one's personal value and using it as a gift, not as a source of pride. In that sense, there's no place for our vanity in God's demanding that we join with Him as partners in the task of managing His world. That is simply the nature of existence.

There is an ancient tradition that, at the end of an individual's earthly life, the question asked at the "Pearly Gates" will not be, "Why didn't you achieve the level of Moses?" but rather, "Why didn't you achieve your own person potential?" We humans are partners with God in running this world. This is not one option among many. It is our obligation. Be fruitful, learn to control nature (Gen. 1:28). Fill with good the lacunae left by the *tzimtzum* of creation. Fix this less than perfect world that we have inherited. Each person, each community, each generation can only act within the potential of its time and environment.

But to use our potential most effectively we have to abandon, actually sacrifice, the popular though erroneous image of God the Father who controls our every act. The biblical image of God implies that God could indeed control every nuance of our acts and every tinge of our thoughts. But a God that would act out that potential power is not the God of the Bible. As made abundantly clear, the God of the Bible has placed that power in our hands.

Within that window of potential, we choose among the locally and temporally available options.[3] Society of a hundred years ago could hardly be faulted for not discovering the cause and the set of cures to the highly variegated plague of cancer. That option was not within the capabilities of a century ago. Today, with our technology and knowledge of genetics and pharmaceuticals, the cure may be within our reach.

In Hebrew, repair of the world is termed *te'kun olam*. When *te'kun olam* is finally made the driving societal priority, then not only will it be "good" individually, but more than that, a harmony will well up between the personal desires of the *nefesh* and the realized Oneness of the *neshama*. That will be "very good," the ultimate *te'kun*.

APPENDIX

With Wisdom God Created the Heavens and the Earth

The Universe as the Symbol of a Thought

Presented at the Smithsonian Institution Conference: Complexity Theory and Semiotics: Unraveling the Mystery of Nature and the Nature of Mystery

May 11, 2002

We live in a world steeped in what George Gilder refers to as the materialist superstition. If we can't see it, weigh it, touch it, it's not there. This view is not so surprising. We ourselves are material beings, in that we are made of matter. All our natural senses respond to matter in one form or another. Over the millennia of our development, it was the material environment that shaped and sharpened our senses.

But a revolution has occurred in this perception. It started with Einstein's amazing laws of relativity, that the passage of time is not constant, and the dimensions of space are flexible. Then came the discovery of the uncertain, fuzzy world of the quantum. And suddenly our classical view of reality, the inbred misconception that reality must conform to our logic, was shattered. We have discovered that the reality we perceive stands in place of, or better said, represents, a deeper essence of truth.

And that is what I discuss here, the idea, admittedly speculative, that the truth of our universe is not as we perceive it, even with the aid of the most sophisticated particle accelerators and most powerful space telescope; that from the invisible realm of the quantum to the vast reaches of space, our universe may more closely resemble a thought than a thing.

The study of our universe may be the ultimate exercise in semiotics.

THE PHYSICAL UNIVERSE is real, real in the sense that it's tangibly out there. I measure the size, weight, hardness of an object. Get values. Other persons do the same and they get the same data. It's not my imagination that when I look out from my porch, I see the row of cypress trees planted there to mark our property line. I'm part of the physical world and so are the trees.

And therein lies the rub. We are all part of the same system. Is that perception made by me of a solid material world an artifact of how I, the perceiver, am built? I believe it was Bertrand Russell who said: The idea that there are hard little lumps that are electrons, protons, neutrons, is . . . derived from our perception of touch. We perceive the world as particulate because the personal encounters we have with the world are primarily tactile. But it is an error to confuse our perception of reality with reality itself.

We are so deeply and totally within the system that we view it in self-referenced terms. I mean if ice could speak and one lump of ice touched another, would it say 'my what a cold lump you are?' No. It would see it as being part of the same. I wonder if fish are any more aware of the water within which they swim than we are aware of the continuous stream of personal self-consciousness, the "I" of the self that nonstop floods our heads.

But when we view our world from a more precise perspective, it is a very, very different view that we get. Just as with a photo in the newspaper, up close it is only a mass of dots and spaces.

On the micro-scale, those things we call atoms that join together to become the solids we know and feel. The positive nucleus surrounded by the negative electrons. Pump up the nucleus to the size of an orange or grapefruit and where's the electron cloud? Four miles out in each direction. Four miles of exquisitely empty space. Not a space filled with air. Air is stuff. The four miles would be exquisitely empty of everything, filled only with virtual never-seen imagined photons that somehow bind the electrons in their bands of orbits. That volume ratio of an orange to a sphere four or so miles in diameter is one part in 10^{15}. Imagine the impossibility of a task to find a single orange within a sphere four miles in radius. It could take a lifetime. Solid though a stone may feel, it is really almost entirely empty space made to feel solid by virtual, never-seen forces.

And in the micro-pico world of the quantum, even the protons and neutrons and electrons fade away into a fuzzy haze, a cloud of forces.

It was Louis de Broglie, in the 1920s, who opened a Pandora's box with his insight, first as theory and then as experiment, that matter as well as light must possess wavelike properties. With this realization, particles became waves, extended, no longer definite in size.

I wonder, as I look at the finger-shaped leaves of those cypress trees, just what aspect of nature is reality and what is the metaphor.

Few scientists today argue for a universe without the big bang. The standard model is that somehow, from absolute nothing—nothing in the sense of not a thing, not material or time or space as we know them—came a massive burst of exquisitely hot energy, electromagnetic radiation, or, in other words, superpowerful light beams. That was the beginning and everything that exists or has existed was formed from that initial energy. We are made of and are powered by that fifteen-billion-year-old burst of energy.

Now I have no problem in understanding how a craftsperson might turn an amorphous lump of silver into a beautiful bowl, but I do not have a clue as to how a burst of energy, akin to superpowerful light rays having no mass whatsoever, can metamorphose and become the solid elements that combined to form all the material world. Yet we have discovered that the entire universe is the manifestation of the energy of the big-bang creation, articulated in a myriad of different forms.

René Magritte's painting of a pipe looks exactly like a pipe, and from a distance it seems real enough to smoke. But up close we see it's just paint on canvas. What would our world look like if we could view it really up close?

"Tradition can be a parasite, even an enemy." So wrote Frank Lloyd Wright in his brilliant classic *The Natural House.* Our brain is surfaced by the cortex, seat of our pure logic. But under the cortex lies the limbic system, filled with emotion and memories. And those memories strongly shape how we handle our logic. At times we cling to our traditions even when our logic tells us they are counterproductive and even wrong.

The song a sparrow learns in its youth is its song for life. And we humans are no different. I learned that the atom is made of hard little nuggets called protons and neutrons. And now we have discovered something very different. But it is hard to replace the song of youth. Illogical though it seems, those imagined particles of the subatomic world have turned out to be fields of force, fuzzy and extended. Could it be that there is a reality even deeper than those forces, a single substrate from which everything flows?

KNOWING THE STRUCTURE of a water molecule, the 104° bonding angle formed by the two hydrogen atoms as they each share their electron with the single oxygen, we can predict that high-energy H_2O is gaseous, with no fixed order among the molecules. We call it steam or vapor. Moderate-energy H_2O becomes somewhat more organized, forming a liquid. Low-energy H_2O forms the ice crystal, a model of organization. This is all intrinsic in the chemistry and physics of the H and O atoms, the sharing of their electrons. I could predict the existence of water and ice and steam, all that from the basic laws of the chemistry of oxygen and hydrogen. A totally reductionist approach, even if I had never seen hydrogen or oxygen gases or water.

But now step back a few stages, to a time before the existence of H's and O's, before atoms, and before quark confinement, to the moment of the big-bang creation when all the world was composed of energy. As space stretched out and energy levels fell, a tiny part of that energy changed form and became solid, protons and neutrons, and finally you and me. How? Intrinsic in the H_2O molecules are the expressions of gas, liquid, solid. That is built into the mass and charges of the atoms.

Is there something intrinsic in the wave/particles of the big-bang energy that yields the sensation of solidity when they reach a certain level? Is there something we don't know about radiation, its nature or structure, that lets energy assume the form of matter?

I'll take a reductionist approach and look at the universe from the beginning to see what we can learn from "first principles." Even with such a simplistic method, we'll find a universe very different from that which we perceive even with our unaided senses.

I'll assume that we know all the laws of nature, a cookbook of the physics and chemistry of the universe. The first required caveat would be that for some bizarre reason, the self-annihilation of particle/antiparticle pairs would not be complete. As the energy of the big-bang creation condenses, it forms matter and antimatter theoretically in equal amounts. This could lead to total annihilation of all solid matter. Such was not the case. A tiny fraction of the matter survived and we are here as living evidence of that reality, as is every other tangible part of our magnificent universe. With this in place, then based on the laws of nature and the initial conditions of the universe, I could predict that through the alchemy of stellar temperatures and the immense pressures of supernova, the ninety-two stable elements would form. I'd know that among those elements would be sodium and chlorine. I could predict that they could chemically react, forming sodium chloride, common salt. All that would be known from first principles.

But could I predict that in some marvelous combination of the building blocks of matter I'd find joy, sentience, awareness of emotions, the metaphysical flight of love? Not likely. In one mix of protons, neutrons, and electrons, I get a grain of sand. I take the same protons, neutrons, and electrons, put them together in a different mix, and get a brain that can record facts, produce emotions, and from which emerges a mind that integrates those facts and emotions and experiences that integration. It's the same protons, neutrons, and electrons. They had no face-lift, yet one seems passive while the other is dynamically alive.

Nowhere in the brain is the bright green of a leaf, the blue of the sky. But I see the green leaf and marvel at the beauty of the sky. I hear sound but there is no sound in my brain. From where does all this replay of my senses arise, a replay that seems as if it were physically there in my brain? If it is, it is very well hidden.

The facile answer is that we interpret the biochemistry of the brain's auditory and visual systems as sound and sight. Of course, that is the case, but where?

A jumble of letters has no meaning and in the letters there is no hint of the arbitrary meaning of a word. But from them, when joined together according to rules of a language, a sonnet can emerge. A blank surface and a pail of paint tell nothing, but by skilled combination a picture emerges so powerful that one touches the canvas to see if it is real. The sonnet is not in the letters any more than the painting is in the paint. But for physical articulation, the sonnet needs the letters and the picture needs the paint. The charge of an electron emerges from an electron but the charge is not made of the electron. Does mind, sentience, emerge from the brain in a similar manner? Every level of existence, from the crystalline structure of salt to the changing complexity of a brain, is built from among the same ninety-two elements of our universe that in turn are made of a mix of protons, neutrons, and electrons. This being the case, at what level of atomic complexity does sentience, awareness, emerge?

Freeman Dyson, a physicist at the Institute for Advanced Study, Princeton, avers that it enters at a very basic level: "Atoms are weird stuff, behaving like active agents rather than inert substances. . . . It appears that mind as manifested by the capacity to make choices is to some extent inherent in every atom." Can mind be a part of an inert matter, an atom?

John Archibald Wheeler, former president of the American Physical Society, physics professor emeritus of Princeton University, winner of the Einstein Award, gives a clue to how that might be. Wheeler sees the world as the "it" (the tangible item) that came from a "bit" (eight of which comprise a byte of information). He is quoted as having first viewed reality as being composed of particles. Then as his understanding broadened, the particles were seen not to be particles at all, but rather the manifestation of fields. Now after a lifetime of study, reality appears to be the expression of information.

Shoucheng Zhang at Stanford, Anton Zeilinger at the University of Vienna, and Ed Fredkin at M.I.T. voice the same speculation: that matter actually arises from a structured or organized substrate of information. That tangible matter is actually the manifestation of ethereal information.

It all sounds bizarre. And it is nothing like the song I learned in my youth. Yet it is only slightly more outrageous than the proven phenomenon of intangible energy metamorphosing into matter.

The late George Wald, Nobel laureate, organic chemist, professor of biology at Harvard University, envisioned that mind is the source of matter. This makes all the sense in the world if, in fact, matter is built from energy and energy is built from information. Suddenly, the old conundrum of how the physical brain gives rise to the ethereal mind and experienced sentience evaporates. It is not a question of consciousness arising from matter. It is rather quite the opposite, of matter arising from consciousness.

MIND, AS INFORMATION or wisdom, is present in every atom. Mind is ubiquitous in our universe, just as wisdom is the basis of all existence.

The tree and every other part of nature express in physical form the wavelike ethereal energy from which they are fashioned. And that elemental energy is none other than the manifestation of the wisdom from which it is built. The existence so familiar to our human sense of touch is but a metaphor that subtly implies an underlying truth far grander in its simplicity than that of the most exotic complexity of life and brain. The diversity of the cosmos, built of time and space and matter, has arisen from a singularity, not of the physical type couched within a black hole, but of a unity brought into being as mind, the first act of the creation. And the wisdom of our minds, if used properly, can close the loop, linking with this underlying mind of the creation.

J. A. Wheeler, during a BBC special, "The Creation of the Universe," summarized the quest for ultimate reality: "To my mind, there must be at the bottom of it all, not an utterly simple equation, but an utterly simple IDEA. And to me that idea, when we finally discover it, will be so compelling, and so inevitable, so beautiful, we will say to each other, 'How could it have ever been otherwise?' "

Plato described our perception of life as if we were persons viewing shadows on a wall, totally unaware of the reality that produced those two-dimensional images. The prophet Isaiah, three hundred years prior to Plato, laid the basis for Plato's analogy:

> *Then the eyes of the blind shall be opened and the ears of the deaf unstopped. . . . The people that walked in darkness have seen a great light; they that dwelled in the land of images, upon them the light has shown.* (Isaiah 35:5; 9:1)

The study of our universe turns out to be an exercise in semiotics.

Acknowledgments

Moses Maimonides, the twelfth-century theologian and philosopher, wrote of society being a web of individual efforts. Be thankful, he told us, to those who prepare our daily portions of bread. If it were not for their efforts, each of us would need to use all our time planting and plowing, nurturing and harvesting the wheat. Then winnowing and grinding, sifting and baking. There'd be scant time to further any of the activities we deem so essential to our high-tech, advanced medical civilization. The writing of a book is not so very different. A variety of talents contribute. The author is merely the person who brings the weave together, what is termed in Talmudic literature as *ma'keh b'patish,* the final tap of a hammer to complete the work.

In composing *God According to God*, I am beholden to many persons, most significantly, my wife, the author Barbara Sofer, the patient, brilliant mother of our children, Avi, Josh, Hadas, Yael, and Hanni, all of whose discussions contributed to the thoughts herein. Meetings with Roy Abraham Varghese and Antony Flew yielded highly valuable material. My many talks with Gil Goller, Dennis Turner of blessed memory, Dennis

Prager, and Barry Kibrick ("Between the Lines") reinforced many of the ideas incorporated here. Among others were Avraham Rosental; Carl Hunt; physics professor par excellence Robley D. Evans, Rabbi Noah Weinberg, and Zola Levitt of blessed memory; Sandra Levitt; Franklin Wong; Lexie Coxon; Uriel Simon; Yaacov Fogelman; Daniel Reisel; Richard Dawkins (whose criticism during a vigorous debate helped to sharpen my response to atheists); and Rabbis Herman Pollack of blessed memory, Heschel Weiner, Chaim Brovender, Shlomo Riskin, David Lapin, and Daniel Lapin.

In the formal presentation of *God According to God*, four persons were central. Cynthia (Cindy) DiTiberio's wise editing kept the theme in focus, sifting the desired wheat from the overly abundant chaff. In the final copyediting, Alison Petersen, Suzanne Stradley, and Ann Moru helped form the final text into a finished work.

To all I extend my warm thanks. And of course I acknowledge my reliance upon the Force ever active behind the "proscenium" upon which we live out our lives.

Notes

CHAPTER 1: A FEW WORDS ABOUT WHAT GOD IS NOT

1. Bart Ehrman, *God's Problem: How the Bible Fails to Answer Our Most Important Question—Why We Suffer* (San Francisco: HarperOne, 2008).
2. Richard Dawkins, *Religion: The Root of All Evil?* BBC Documentary; Channel 4, January 2006.
3. E. O. Wilson, *Consilience* (New York: Alfred A. Knopf, 1998), p. 8.
4. In view of that prohibition, it is not so surprising that Israel is the only country that ended the twentieth century with vastly more trees than it had at the beginning of that century. Millions were planted, sponsored by Israel's national forestry fund.
5. Norman Geisler, *Miracles and the Modern Mind* (Grand Rapids, MI: Baker Book House, 1992).
6. Harold Morowitz, *Energy Flow in Biology* (New York: Academic, 1969).
7. Paul Davies, *The Mind of God* (New York: Simon & Schuster, 1992).
8. Simon Conway Morris, *Life's Solutions* (Cambridge: Cambridge University Press, 2003), p. 328.
9. The word "acquired" is the translation of the Hebrew word *k'na'nie.* In modern Hebrew that is the meaning of *k'na'nie.* In biblical Hebrew, *k'na'nie* has two equal meanings: "acquired" and also "created." From the continuing verses of this Proverb, it is clear that "created" is also deeply within the intended meaning here.

10. James Jeans, *The Mysterious Universe* (Cambridge: Cambridge University Press, 1931).
11. Michael Shermer, "Mustangs, Monists and Meaning," *Scientific American*, September 2004.
12. For an insightful discussion of what "Adam's rib" might actually have been, I urge those interested to read the medically significant report, written by a professor of developmental biology and a professor of biblical literature, that first appeared in the prestigious *American Journal of Medical Genetics* in 2001: "Congenital Human Baculum Deficiency," by Scott F. Gilbert and Ziony Zevit.

CHAPTER 2: THE ORIGINS OF LIFE

1. Steven Weinberg, "A Designer Universe?" in Paul Kurtz, ed., *Science and Religion: Are They Compatible?* (Amherst, NY: Prometheus, 2003), p. 38.
2. The concept of the universe being the result of a vacuum fluctuation was first suggested by Edward Tryon in 1973 in an article titled "Is the Universe a Vacuum Fluctuation?" Though published in the prestigious peer-reviewed journal *Nature* (vol. 246), Tryon's insight was ignored for a decade and a half until the discoveries of cosmology blended with what was until then considered to be quite an extreme proposition.
3. For an almost reader-friendly description of how the world came to be via quantum physics, I recommend "The Last Great Mystery," *Discover* magazine, April 2002.
4. In a bit of irony, 1953 was also the year that the scientists Francis Crick, James Watson, and Maurice Wilkins reported on their discovery of the unique structure of the genetic code, DNA (deoxyribonucleic acid).
5. Katie Edwards and Brian Rosen, *From the Beginning* (London: Natural History Museum, 2004), p. 13.
6. Stephen Hawking, *A Brief History of Time* (New York: Bantam Books, 1998), p. 123.
7. R. Kerr, "Did Darwin Get It All Right?" *Science* 267 (1995): 1421.
8. Max Tegmark, "Parallel Universes," *Scientific American*, May 2003, p. 35.
9. In Hebrew two words indicate "order." *Boker*, as pointed out by Nahmanides, means "able to be discerned." *Seder* relates to a series of events or objects set in a fixed order; the Passover holiday follows a *seder*, an order, in the telling of the Exodus drama on Passover evening.
10. S. Kumar and S. Subramanian, "Mutation Rates in Mammalian Genomes," *Proceedings of the National Academy of Science* 99, no. 2 (January 22, 2002).
11. D. Voet and J. Voet, *Biochemistry*, 3d ed. (New York: Wiley, 2004), p. 279.
12. Jon Seger, "How Can an Organ as Complex as an Eye Evolve?," *Scientific American*, March 2008, p. 87.
13. Simon Conway Morris, *Life's Solutions* (Cambridge: Cambridge University Press, 2003), p. 309.
14. Conway Morris, *Life's Solutions*, p. xv.
15. George Wald, "Life and Mind in the Universe," Quantum Biology Symposium, *International Journal of Quantum Chemistry* 11 (1984): 1–15.
16. George Wald, "The Origins of Life," *Scientific American*, August 1954, p. 48.

17. Clair Folsome, "Life: Origin and Evolution," *Scientific American*, special publication, 1979, p. 45.

CHAPTER 3: THE UNLIKELY PLANET EARTH

1. Fred Hoyle, "The Universe: Past and Present Reflections," *Engineering and Science*, November 1981.
2. Antony Flew, *There Is a God: How the World's Most Notorious Atheist Changed His Mind* (San Francisco: HarperOne, 2007).
3. For a reader-friendly but not scientifically condescending description of the unique location and size of the earth, see Katie Edwards and Brian Rosen, *From the Beginning* (London: Natural History Museum, 2004).

CHAPTER 4: NATURE REBELS

1. Christian de Duve, *Tour of a Living Cell* (New York: Scientific American Books, 1984).
2. James Jeans, *The Mysterious Universe* (Cambridge: Cambridge University Press, 1931).
3. Virginia Morell, "Minds of Their Own" (*National Geographic,* March 2008), pp. 36–61.
4. Soju Chang, M.D., "Infection with Vancomycin-Resistant *Staphylococcus aureus* Containing the *vanA* Resistance Gene," *New England Journal of Medicine* 348 (2003): 1342.
5. Frank Vertosick, Jr., *The Genius Within* (New York: Harcourt, 2002), p. 53.
6. Freeman Dyson, "Progress in Religion" (acceptance speech, Templeton Prize), March 2000.
7. The Hebrew word for earth is *adamah,* from which is derived the name Adam.
8. For those interested in a more broadly based discussion of the interplay and amazing similarity between the biblical and scientific descriptions of our origins, my first two books, *Genesis and the Big Bang* and *The Science of God,* provide the details.
9. The Tree of Life is also the source of Divine wisdom. We learn this in Proverbs: "She is a Tree of Life to those that hold fast to her" (3:18). And who or what is the "She"? Of that we read earlier: "I am wisdom ['wisdom' in Hebrew, *hok'mah,* is a feminine noun; hence, 'She']. . . . The Eternal God acquired me [wisdom] as the beginning of His way, the first of His works of old" (8:12, 22).

CHAPTER 5: A REPENTANT GOD?

1. It is important to emphasize that regardless of whether nine-hundred-year life spans and the worldwide Flood of the Bible are literal or legendary or metaphorical, there is a central theological lesson to be learned here, and the Bible is teaching it. We have learned that, in a Divinely managed world, there are times when God lets the system get so far off course that a major revamping is needed. We are forced to ask the almost obvious question: Couldn't God have foreseen and therefore averted the fiasco?
2. There is a more modern example of a king and a misdirected architect. In 1538, the sultan of the Ottoman Empire, Suleiman I, decided that Jerusalem required

a refurbished surrounding wall. But sultans do not build walls. They hire architects. Two were employed to design and oversee the wall's construction. When the sultan came to view the completed structure, he was astounded to discover that the architects had erred. They had excluded the south and southeast portions of Mount Zion from the enclosure. The architects were to be executed for their blunder. They were granted one request prior to death. Their final wish was to be buried inside the walls of the city. And they were. Their tombs lie a few meters past the entrance to the Old City of Jerusalem, through the Jaffa Gate, in a niche on the left.

CHAPTER 6: ARGUING WITH GOD

1. The argument used by Moses to challenge God's plan is instructive of the role of Israel in the world. Israel is not chosen because it, as a people, is alone important to God. If that were the case, then what would it matter if the Egyptians thought wrongly of God's intent? Moses uses a similar argument several months later when the spies bring back a bad report about the land of Canaan, urging the people to return to Egypt (Exod. 13): "And the Eternal said to Moses, 'How long will this people despise Me? . . . I will strike them with pestilence and destroy them. . . .' [And Moses said,] 'If you kill this people as one person, then the nations which have heard of Your fame will speak saying: "Because the Eternal was not able to bring them into the land which He promised to them, therefore He has slain them in the wilderness"'" (Num. 14:11–12, 15–16). Again, if the people of Israel were the only concern of God, what the other nations thought would be irrelevant. In fact, the Divine goal is for all peoples to discover the one Creator and that Creator's active role in history. The chosenness of Israel and its obligations as listed in the Torah are designed to set Israel visibly apart from all other nations. By that separation, Israel's flow through history is readily observed. This flow is bizarre. It is in fact unique. For many it demonstrates God's active presence in history, directing that flow. The first of the Ten Commandments might have recalled the overwhelming power of God to create the universe. Instead, the first commandment relates to God active in history: "I am the Eternal your God who has taken you out of Egypt, from the house of slavery" (Exod. 20:2).

   The idea of Israel being a "holy people" matches totally its role in history, but not according to the modern definition of "holy." The Hebrew word translated "holy" is *kodesh,* which does not mean "exalted" or "wonderful," as might be erroneously inferred from modern usage of the word. Biblically, *kodesh* means "separate," "set apart." Israel is a people set apart, a visible indicator that indeed God interacts in this world. History has proven the accuracy of the word.
2. Some aspects of the argument that I present here, that God expected Abraham to argue and when Abraham failed to argue, God altered their mutual relationship, have also been given in part by other students of the Bible. Among those are Alan Dershowitz in his book *The Genesis of Justice* (New York, Warner Books; 2000); Shulamit Aloni, former Minister of Education in Israel, in "So Who Was Testing Whom?" *Jerusalem Post,* October 3, 2003; and the noted French philosopher Emmanuel Levinas in his *Totality and Infinity* (Pittsburgh, PA; Duquesne University Press, translated by Alphonso Lingis, 1969).

CHAPTER 8: LIFE AND DEATH

1. Pim van Lommel, et al, "Near-Death Experience in Survivors of Cardiac Arrest," *Lancet* (December 2001): 2039–2045.
2. Werner Heisenberg, *Physics and Beyond* (New York: Harper & Row, 1971).
3. Robert Sapolsky, "Bugs in the Brain," *Scientific American,* March 2003, p. 71.
4. John Maddox, "What the Next 50 Years Will Bring," *Scientific American,* December 1999.
5. Erwin Schroedinger, *My View of the World* (Cambridge: Cambridge University Press, 1964).

CHAPTER 9: THE DESERT TABERNACLE

1. Rabbi Tanhuma, ca. 400; based on Leviticus 8:3–4.
2. That figure, 100,000 light-years, is the distance light would travel in 100,000 years. The speed of light is 300,000 kilometers a second. Clearly 100,000 light-years is a huge distance. The initial burst of expansion of the universe was extraordinary. The expansion still continues, though at a more "leisurely" rate.
3. Talmud, *Brachot* (*Blessings*), 55A.
4. The Hebrew alphabet (aleph, bet, gimmel, daled . . .) is estimated to predate the Greek (alpha, beta, gamma, delta . . .) by centuries, and so would form the earlier base of the English *A, B, C, D.*
5. Kenneth C. Davis, *America's Hidden History* (San Francisco: HarperCollins, 2008).
6. Christopher Hitchens, *God Is Not Great: How Religion Poisons Everything* (Twelve Publishing, 2007).

CHAPTER 10: KNOWING TRUTH IN YOUR HEART

1. Talmud *Yadiim* 3:5. The word Yadiim means "a pair of hands" and contains the root of the Hebrew word for friendship, a "holding of hands."

CHAPTER 11: UNDERSTANDING THE MERCIFUL GOD OF THE BIBLE

1. Just as with the Flood at the time of Noah, don't get stuck here on whether it was an actual fish, or if the tale is myth or metaphor. The message of this chronicle is the teaching found in the juxtaposition of the strict, initially quite human response of Jonah with the more lenient response of God.
2. The final phrase, "and also much cattle," added almost as an afterthought, comes to inform that the compassion of the Creator extends to animals as well as to humans.

CHAPTER 12: PARTNERS WITH GOD

1. Steven Weinberg, *Dreams of a Final Theory* (New York: Pantheon, 1993), p. 244.
2. Joseph A. Hertz, *Commentary on the Torah* (London: Soncino Press, 1993), Deut. 30:19.
3. In the development of life, Charles Darwin also envisioned a window of potential. His *Origin of Species* details the driving force of evolution, that of the survival of the fittest. Yet in the closing lines (including in the final edition published during his lifetime, 1872), Darwin wrote, "There is a grandeur in this view of life, with

its several powers, having been originally breathed by the Creator into a few forms or one." (Unfortunately, when quoting Darwin this phrase acknowledging the role of "the Creator" is usually replaced by "...," that is, it is omitted.)

That life developed from the simple to the complex is indicated by a variety of sources. This is the account brought in the opening chapter of Genesis. It is also evidenced within the fossil record and by the discoveries of the genetic similarity of all forms of life. The debate is not about the flow of life. The debate is what drove that flow. Was it a series of random lucky draws by nature, or is there indication of a teleological directive, or perhaps most accurately, a combination of both?

In *The God Delusion* (Bantam Press, 2006), arch-agnostic/atheist Richard Dawkins attributes to luck the key stages in life's flow—that of the origin of life from nonliving matter, and then the emergence of consciousness, self-awareness, from within that life. Dawkins's resort to luck is not a retreat to the-god-of-the-gaps. It is merely an acknowledgment that there is no hint within the prebiotic substrates of life either for life or cognition. Interestingly, as noted earlier, those same two evocative transitions are exactly the two stages that Nobel laureate biologist George Wald attributed to the manifestation of a mind as being the source of all aspects of existence.

We saw that the number of potential blind alleys in protein evolution is so vast that nature must have invented short cuts in order to find the very few protein forms that succeed in maintaining viable life. The secret may lie in an inherent yet subtle ability of nature to self-organize. In a lecture in 2004, Sir John Polkinghorne cited research by Stuart Kauffman related to that subject. A system of 10,000 lightbulbs, in which each bulb may be either on or off, has the potential number of 2 to the power of 10,000 combinations. In the usual base ten numerical system, that number is a one with 3,000 zeros after it, or a billion billion billion continued 333 times! However, if the system is not totally free (just as the laws of chemistry and physics that determine the flow of prebiotic and biotic reactions are not totally free but are highly constrained by the laws of nature), but the state of each bulb is made to be dependent on the state of merely 2 other bulbs, the possible number of combinations plummets to 100. This transition from near chaos to relative order is instantaneous. In his *At Home in the Universe* (Viking, 1995), Kauffman discusses in detail the biotic implications of nature's inherent (that is, built-in) ability to self-organize. If this were protein folding, then from among 100 choices, nature could easily find the few that are viable. Simon Conway Morris suggested that somehow nature has the uncanny ability to find among the vast hyperspace of biological possibilities the very few combinations that have proven to be successful. The phenomenon of self-organization, built within the created laws of nature, may be part of that uncanny ability of nature. In that sense, life would indeed be written into the very fabric of the universe.

Scripture Index

HEBREW BIBLE

Genesis

Subject Index